URBAN CLIMATE LAW

COLUMBIA UNIVERSITY EARTH INSTITUTE
SUSTAINABILITY PRIMERS

COLUMBIA UNIVERSITY EARTH INSTITUTE SUSTAINABILITY PRIMERS

The Earth Institute (EI) at Columbia University is dedicated to innovative research and education to support the emerging field of sustainability. The Columbia University Earth Institute Sustainability Primers series, published in collaboration with Columbia University Press, offers short, solutions-oriented texts for teachers and professionals that open up enlightened conversations and inform important policy debates about how to use natural science, social science, resource management, and economics to solve some of our planet's most pressing concerns, from climate change to food security. The EI Primers are brief and provocative, intended to inform and inspire a new, more sustainable generation.

Climate Change Adaptation: An Earth Institute Sustainability Primer, Lisa Dale

Managing Environmental Conflict: An Earth Institute Sustainability Primer, Joshua D. Fisher

Sustainable Food Production: An Earth Institute Sustainability Primer, Shahid Naeem, Suzanne Lipton, and Tiff van Huysen

Climate Change Science: A Primer for Sustainable Development, John C. Mutter

Renewable Energy: A Primer for the Twenty-First Century, Bruce Usher

URBAN CLIMATE LAW

AN EARTH INSTITUTE SUSTAINABILITY PRIMER

MICHAEL BURGER
AND
AMY E. TURNER

Columbia University Press *New York*

Columbia University Press
Publishers Since 1893
New York Chichester, West Sussex
cup.columbia.edu

Library of Congress Cataloging-in-Publication Data
Names: Burger, Michael, 1974– author. | Turner, Amy E., author. | Columbia Earth Institute, sponsoring body.
Title: Urban climate law : an earth institute sustainability primer / Michael Burger and Amy E. Turner.
Description: New York : Columbia University Press, [2023] | Series: Columbia University Earth Institute sustainability primers | Includes index.
Identifiers: LCCN 2023008751 | ISBN 9780231201346 (hardback) | ISBN 9780231201353 (trade paperback) | ISBN 9780231554053 (ebook)
Subjects: LCSH: Climatic changes—Law and legislation—United States. | City planning and redevelopment law—United States. | Building laws—United States.
Classification: LCC KF3819 .B87 2023 | DDC 344.7304/6—dc23/eng/20230727
LC record available at https://lccn.loc.gov/2023008751

Printed and bound by CPI Group (UK) Ltd, Croydon, CR0 4YY

Cover design: Julia Kushnirsky
Cover photograph: Shutterstock

CONTENTS

A Note on Terminology and Glossary *vii*

Introduction 1

1 Cross-Cutting Legal Concepts 9

2 Equity 20

3 Buildings 39

4 Reducing Transportation-Related Greenhouse Gas Emissions 72

5 Scaling Up Renewable Energy 95

6 Decarbonizing a City's Waste 117

Conclusion 140

Notes *143*

Index *175*

A NOTE ON TERMINOLOGY AND GLOSSARY

This primer uses certain terms interchangeably. While cities are often touted as leaders in mitigating climate change, we recognize that towns, villages, and counties also frequently play this leadership role. What's more, cities, towns, counties, and other municipal forms of government have similar constraints and opportunities available to them.

As a matter of style, we use the words *city*, *municipality*, and *locality* interchangeably to refer to local areas in the United States that have an incorporated, substate form of government, including without limitation cities, towns, counties, and villages.

In addition, *greenhouse gases* include a number of gases in addition to carbon dioxide, the most prevalent greenhouse gas. For simplicity, we use the terms *greenhouse gases* (and its abbreviation *GHGs*) and *carbon* to refer to all greenhouse gases.

Similarly, we use terms such as *carbon mitigation*, *decarbonization*, and *GHG emission reduction* to refer to the reduction of greenhouse gas emissions.

The following terms are used throughout this book.

ACCESSORY DWELLING UNIT: a small residential unit located on the same lot as a single-family home.

ALL-ELECTRIC CONSTRUCTION: a new or renovated building powered by electricity as its sole energy source.

BENCHMARKING: a policy of measuring the energy performance of a building over time and relative to other similar buildings to track changes in energy use and identify opportunities for energy savings.

BIPOC: Black, Indigenous, and people (or person) of color.

BUILDING CODE: a range of requirements that set standards for building construction and major renovations; for example, a building or construction code, an energy or energy conservation code, or a plumbing code, each of which may apply to residential, commercial, or some other subset of buildings.

BUILDING ENVELOPE: the physical separation of the interior from the exterior of a building.

CLEAN ENERGY: energy provided by low-carbon energy sources, including but not limited to renewable energy sources. The definition of *clean energy* is open to debate and for some may include nuclear power.

CLIMATE JUSTICE: a framework and movement that acknowledge that climate change has disproportionately adverse impacts on low-income, BIPOC, and other underprivileged populations, and that seeks to address these inequities.

COMMUNITY CHOICE AGGREGATION (CCA): a program that allows an aggregator, often a local government, to arrange for the purchase of electricity in bulk such that residents may purchase the electricity through the program while the traditional utility provider continues to provide distribution and billing services.

COMPLIANCE PATHWAY: one of two or more options for compliance with a federal, state, or local law, rule, regulation, or other requirement.

CONGESTION PRICING: a road pricing strategy that sets a higher price for driving on a road or in a designated area during times with more traffic.

CORDON PRICING: a congestion pricing strategy that sets variable or fixed charges to drive into a geographic zone of a city.

DE FACTO MANDATE: a standard structured to appear as an incentive but that actually compels behavior as a mandate or requirement would. Where federal or state law preempts certain local requirements, a court may hold an incentive preempted by finding it to be a de facto mandate.

DEMAND RESPONSE: a strategy that encourages or incentivizes electricity customers to reduce their electricity use during periods of high demand.

DENSITY BONUS: a zoning incentive that allows a developer to increase the maximum allowable development or number of units for a parcel of land.

DEREGULATED (ELECTRICITY): describes a state jurisdiction in which a loosening of the regulation of the electricity system has taken place, notably with the generation and distribution functions of electricity service split such that customers may choose a power generator other than the local utility.

DILLON'S RULE: the doctrine that a unit of local government has no more power or authority than that expressly granted to it by the state, along with any implied powers necessary to carry out a grant of authority from the state government.

DIRECT-CURRENT (DC) RAPID CHARGER: an electric vehicle charger that uses direct current and can charge a vehicle with at least sixty miles' worth of distance per twenty minutes of charging.

DISTRIBUTED ENERGY GENERATION: the generation of energy in small quantities at or near where it will be used (e.g., rooftop solar panels).

DORMANT COMMERCE CLAUSE (DCC): an implied, or dormant, aspect of the Commerce Clause of the U.S. Constitution that bars state and local governments from passing laws that impermissibly discriminate against or burden interstate commerce.[1]

ELECTRIC READINESS: a construction standard requiring a building to be wired for an electric furnace, HVAC system, and other appliances, even if fossil fuel–powered systems or appliances are installed at the time of construction.

ELECTRIC VEHICLE (EV): a vehicle powered by one or more electric motors.

ELECTRIFICATION (BUILDING ELECTRIFICATION): switching out fossil fuel–powered building systems and appliances in favor of electricity-powered systems and appliances. Also called "beneficial electrification."

EMBODIED CARBON: the cumulative greenhouse gas emissions attributable to the supply chain, transportation, manufacturing, and end-of-life processing of a building's construction materials.

EMINENT DOMAIN: the power of a government to take private property for public use. Per the Fifth Amendment of the U.S. Constitution, a taking by eminent domain requires the government to pay "just compensation" to the property owner.

ENERGY-ALIGNED LEASE: a lease that realigns the incentives between landlord and tenant to better catalyze energy improvements to a building. Also referred to as a "green lease."

ENERGY EFFICIENCY: a strategy or technology to use less energy to perform the same function.

ENERGY JUSTICE: a framework and movement with "the goal of achieving equity in both the social and economic participation in the energy system, while also remediating social, economic, and health burdens on those disproportionately harmed by the energy system."[2]

ENVIRONMENTAL JUSTICE: a framework and movement that prioritize the equal development and protection of environmental laws for all people regardless of race or income level and that seek to prevent or redress inequitable disparities in exposure to environmental pollution experienced by BIPOC and low-income populations.[3] An *environmental justice community* is a neighborhood where environmental and socioeconomic factors contribute to health disparities and other negative outcomes experienced by people of color and those with low income.

ENVIRONMENTAL REVIEW: the process of identifying and assessing the environmental impacts of a governmental project or action, or of a private party's project or action that requires a governmental permit or funding, pursuant to federal, state, or local law.

EQUAL PROTECTION: an individual right under the Fourteenth Amendment of the U.S. Constitution to equal treatment by the government in comparison to the treatment of other people or classes in similar circumstances.

FEDERAL-AID HIGHWAY: "a public highway eligible for [federal funding] other than a highway functionally classified as a local road or rural minor collector."[4]

FEE: a charge or payment for services.

FRANCHISE AGREEMENT: a contract through which a local government offers access to the public right-of-way for a utility to place pipes, wires, and other infrastructure, often in exchange for a fee.

FRONTLINE COMMUNITIES: neighborhoods that experience the earliest and worst impacts of climate change.

FUEL ECONOMY: "the average number of miles traveled by an automobile for each gallon of gasoline (or equivalent amount of other fuel) used."[5]

GENTRIFICATION: the process by which a neighborhood changes because of an influx of more affluent residents.

GREEN LEASE: see *energy-aligned lease.*

GREEN TARIFF: an electric utility offering or rate that allows customers (sometimes only large customers) to source electricity from new renewable energy sources.

GREENHOUSE GAS (GHG): a gas that traps heat in the air. Greenhouse gases include carbon dioxide, fluorinated gases (hydrofluorocarbons and perfluorocarbons), methane, nitrous oxide, and sulfur hexafluoride.

HEAT PUMP: a heating and cooling device that can draw heat into a building to heat it or draw heat from a building to cool it. Heat pumps may use a variety of fuels, but in the context of reducing greenhouse gas emissions, they are often fueled by electricity or geothermal energy (energy derived from hot water or steam drawn from geothermal reservoirs in the subsurface of the earth).

HEAVY-DUTY VEHICLE: a vehicle weighing more than an amount defined by the applicable regulator. The U.S. Federal Highway Administration defines heavy-duty vehicles as those greater than 26,000 pounds, whereas the U.S. Environmental Protection Agency defines them as vehicles greater than 8,500 pounds.

HVAC: heating, ventilation, and air conditioning.

HYBRID VEHICLE: a vehicle powered by both an internal combustion engine (i.e., gasoline or diesel) and an electric motor.

INDEPENDENT SYSTEM OPERATOR: an independent entity that coordinates the regional transmission of electricity and ensures the safety and reliability of a region's electricity system.

INVESTOR-OWNED UTILITY (IOU): a privately owned electric utility business subject to state regulation.

JUST COMPENSATION: a governmental entity's payment to a property owner for property it has taken under eminent domain or through a regulatory taking.

LAND USE: the field relating to planning and regulating the development of real estate.

LOCAL (OR LOCALIZED) AIR POLLUTION: emissions into the air of carbon monoxide, nitrogen oxides, sulfur oxides, particulate matter, and other pollutants with significant impacts on the geographic area where they are emitted.

LOW-EMISSION ZONE (LEZ): a bounded geographic area in which certain vehicles are restricted or disincentivized from entering otherwise public roads.

MARKET PARTICIPANT EXCEPTION: the principle that a state or local government does not violate the Dormant Commerce Clause by acting as a buyer or seller in the market. With respect to federal statutes such as the Clean Air Act, the Energy Policy and Conservation Act, and the Federal Aviation Administration Authorization Act, *market participant exception* refers to an exception from preemption under those statutes when the state or local government acts as a market participant rather than a regulator.

MEGAWATT: a unit of measurement equal to one million watts of electricity.

MEGAWATT-HOUR (MWH): one megawatt of electricity generated consistently for one hour.

MICROGRID: a local energy grid that can disconnect from the traditional grid and operate autonomously to provide resilience in the event of grid disruptions.

MODE SHIFT: a change in form of transportation, as from driving to public transit.

MUNICIPAL HOME RULE: a state legislative delegation of broad authority to a local government to govern with respect to its own affairs and other local matters. Also called "home rule."

MUNICIPAL UTILITY: a utility owned by a unit of local government.

MUNICIPALIZATION: the process of forming a new public utility, particularly an electric or energy utility, such that utility assets are owned and electricity or energy is provided by a unit of local government.

NATURAL GAS BAN: a policy that prohibits or restricts natural gas connections to new buildings.

NET METERING: a metering and billing arrangement that allows owners of distributed energy generation systems (such as building owners with rooftop solar panels) to be compensated for energy they transmit to the electricity grid.

NONDELEGATION DOCTRINE: the principle limiting the ability of one branch of government, such as a legislative body, to transfer its authority to another governmental branch or a third party.

ORDINANCE: a local law or regulation.

ORGANIC WASTE: waste derived from living organisms, particularly food waste and yard waste.

"PAY AS YOU THROW": a system of municipal solid waste collection in which residents are charged based on the amount they dispose of.

PENALTY: a fine assessed for violating a law or regulation.

PERFORMANCE STANDARD or **PERFORMANCE PATHWAY:** a requirement, or a compliance pathway within a requirement, that sets a standard of performance for a building but does not specify how that standard must be met.

POLICE POWER: a state's Tenth Amendment right to enact and enforce laws protecting the public's health, safety, and general welfare. A state may delegate all or portions of its police power to local governments.

POWER PURCHASE AGREEMENT (PPA): a contract for energy between an energy project developer and a customer that sets a price for such energy and a time frame for the agreement.

PREEMPTION: the superseding or invalidating of the law of a lower jurisdiction by the law of a higher jurisdiction; the principle that higher levels of law have primacy over lower levels of law. Federal law can preempt state and local law; state law can preempt local law.

PRESCRIPTIVE REQUIREMENT or **PRESCRIPTIVE PATHWAY:** a requirement, or a compliance pathway within a requirement, that specifies actions and items that must be completed to bring a building into compliance.

PRIVATE RIGHT OF ACTION: the right of a private party to bring a case.

PROCUREMENT: the process by which a unit of government purchases or contracts for goods and services.

PROPERTY-ASSESSED CLEAN ENERGY (PACE) FINANCING: a financing mechanism by which property owners can pay back loans for building energy efficiency and renewable energy projects through a voluntary tax assessment tied to the property rather than the borrower.

PUBLIC TRUST: the principle that the government maintains, and is responsible for protecting the public's right to use, certain resources for the public's use.

PUBLIC UTILITY (OR SERVICE) COMMISSION: a state-level agency that regulates public utilities.

PUBLIC UTILITY (OR SERVICE) LAW: a state statute or set of state statutory provisions, and any regulations promulgated thereunder, that regulate public utilities and the provision of energy services.

RATE-MAKING: the process by which a public utility commission sets electricity rates.

REACH CODE: a local building energy code more stringent than the statewide base building energy code.

REDLINING: a practice by banks and federal agencies to deny mortgages for homes in predominantly Black or other BIPOC neighborhoods. While unlawful, redlining was common during several decades of the twentieth century, and its consequences persist today.

REGIONAL TRANSMISSION ORGANIZATION: an entity similar to an independent system operator but generally operating in a larger region.

REGULATION: a rule or order promulgated by an administrative agency. Regulations generally have the force of law.

REGULATORY TAKING: a governmental requirement that restricts the use of private property such that the property owner is deprived of all "economically viable use of his land."[6]

RENEWABLE ENERGY: energy derived from sources that are naturally replenishing but limited in flow, including wind, solar, geothermal, hydro, and tidal energy.

RENEWABLE ENERGY CERTIFICATE OR CREDIT (REC): a market-based instrument that represents the property rights to the environmental attributes of one megawatt-hour of renewable electricity generation.

RENEWABLE PORTFOLIO STANDARD: a regulatory requirement, often by a state, to generate a specified percentage of a jurisdiction's energy from renewable sources.

RESILIENCE: the ability of a community or individual to prevent, withstand, respond to, and recover from disruptive climate impacts.

RETROFIT: a modification to an existing building to make it more energy efficient or improve its energy performance.

SETBACK: the minimum amount of distance required between a lot line and a building line, usually as specified in a zoning ordinance.

SMART METER: a meter that measures and records electricity use at frequent intervals (e.g., hourly or every fifteen minutes) and provides such data to the utility and the customer.

SOLAR READINESS: a construction standard requiring a building to be wired for solar panels, even if panels are not installed at the time of construction.

SPLIT-INCENTIVE PROBLEM: a misalignment of incentives between two parties to a contract, as between a landlord and tenant under a traditional lease, that can ineffectively incentivize both parties to make needed building energy retrofits.

STATUTE: a law enacted by a legislative body.

STRETCH CODE: a building energy code set at the state level that is more stringent than the statewide base building energy code that local governments may adopt if they choose.

TAILPIPE POLLUTANTS: local air pollutants emitted as vehicle exhaust.

TAX: a charge imposed by a government on residents, businesses, transactions, or property to raise revenue.

TRADITIONALLY REGULATED (ELECTRICITY): describes a state jurisdiction in which electric utilities have vertically integrated monopolies providing both generation and distribution, such that electricity customers may not choose a power generator other than the local utility.

UTILITY: a business that provides an essential public service (e.g., electricity or other source of energy) and is therefore subject to regulation as a utility.

UTILITY-SCALE ENERGY GENERATION: large-scale energy-generation projects, often defined as projects of ten megawatts or larger.

VEHICLE-TO-GRID (V2G): a technology by which a plug-in electric vehicle can store energy in its battery and distribute it to the electricity grid during periods of high demand.

VIRTUAL POWER PURCHASE AGREEMENT (VPPA): a "financial agreement in which a customer agrees to pay a predetermined price for the generated electricity and, typically, the renewable attributes (RECs [renewable energy credits]) from a renewable energy project. Instead of the customer receiving the electricity physically, the project owner sells the energy into the local organized wholesale market; for each [megawatt-hour], the buyer then pays or receives the difference between the wholesale market revenue" and the agreed PPA price.[7] Also referred to as a "synthetic" or "financial" PPA.

WASTE-TO-ENERGY: the process by which solid waste is used to produce steam to generate electricity.

ZERO-EMISSION VEHICLE (ZEV): a vehicle that does not emit air pollution from its tailpipe. Electric vehicles fully powered by an electric battery are categorized as ZEVs.

ZERO WASTE: an objective or movement to eliminate landfilled waste by reducing and diverting waste products such that they can be recovered, reused, or recycled.

ZONING: the division of land within a municipality into separate districts with various land use, building size, and other regulations.

URBAN CLIMATE LAW

INTRODUCTION

This book seeks to answer one question: *How can U.S. cities enact and implement promising, ambitious climate mitigation policies that comply with federal and state law?*

Cities have long been at the forefront of climate law and policy innovation. During the years of the Trump administration, cities took on an even more visible role, implementing novel approaches to reducing greenhouse gas (GHG) emissions like mandating net-zero buildings for new construction, scaling up the generation and consumption of renewable energy, and reimagining old systems to encourage vehicle electrification and nonvehicle transportation. Dozens of U.S. cities have now set ambitious climate goals—an 80 percent reduction in GHG emissions, net-zero emissions, and carbon neutrality are all common formulations—and more than 180 have pledged to achieve a 100 percent renewable or carbon-free energy supply.[1] These "climate cities" represent about forty million residents.[2] The Biden administration has ushered in a wholesale change in federal climate policy, but so too has the Supreme Court limited the scope of federal regulatory authority in this space. Regardless of federal and state action on climate, cities' climate commitments remain critical to reducing GHG emissions in line with national and international targets.

Cities may be conceptualized, and imagined, in many ways: as places defined by the aggregated expressions of individuals, families, and communities that reside within a given set of geographical boundaries; as social–economic systems defined by powerful economic interests, such as real estate developers and other large firms and corporations, and the needs of smaller, less powerful, more diffuse businesses and business owners; as particularized iterations of built environments interlaced with urban nature; and as destinations for tourism or migration to name just a few. In this book, the cities we have in mind reflect this diversity of perspectives and interests, as represented by the political leaders, policy entrepreneurs, and expert technocrats within local governments across the United States. These are the actors most closely engaged in climate policy-making: mayors, city councils, climate and sustainability offices, and departments charged with overseeing buildings, streets, energy systems, social support systems, and other infrastructure.

A range of policy options are available to cities, including both new approaches and those that have been effectively deployed in other places. Politics and geography vary, of course, as do the nature and extent of cities' legal authority. Cities are constrained by law, just as they are by technology, finances, existing infrastructure, and social and political realities. They are not free to adopt any and all measures; they cannot simply "cut and paste" climate policies from elsewhere. They must consider the contours of their legally delegated authority, and their policy choices must comport with federal and state law.

Such considerations lead into a thicket of complex legal questions with cross-cutting applications. What is the nature of a city's municipal authority? What is the proper characterization of the legal action at issue? How does it relate to existing state and federal laws? Are constitutional dimensions in play?

These considerations also lead to critical questions of whether and how climate policies may make cities more equitable and just places to live and work. What opportunities exist for centering social equity and environmental justice in urban efforts to reduce GHG emissions and advance the energy transition? What obstacles are in the way?

Typically, a city looks at decarbonization strategies across a constellation of sectors that contribute to a city's total GHG emissions. As described in the *Global Protocol for Community-Scale Greenhouse Gas Emission Inventories* (*GPC*), these sectors include transportation, waste, and stationary energy, a category that includes both energy generation and the buildings that use that energy.[3] In addition, equity and climate justice are implicated in both the disparate impacts of climate change and the policy choices cities make to address them and are key components of decarbonization.

BUILDINGS

Buildings are often a city's first or second largest source of GHG emissions, and because municipalities generally have more authority over local buildings than they do over vehicles, buildings tend to be an early and impactful focal point for a city's plan to reduce GHG emissions. Available tools include building energy benchmarking, improving energy efficiency through retrofits, updating local building codes, restricting or prohibiting the use of certain fuels, requiring or incentivizing green roofs, and promoting green or energy-aligned leasing.

Legal issues that arise in this space include the potential for state preemption of local construction requirements for new and renovated buildings; federal preemption of some energy standards for appliances; privacy protections for the monitoring of home energy use; avoiding improper delegation of building regulatory authority; misaligned incentives in the rental market; and structuring mechanisms like taxes, fees, and penalties. Some of the more novel policies, like building performance standards and all-electric construction requirements, encounter many or all of these considerations. And because buildings are literally where people live, decarbonization laws can interact with other existing laws in ways that affect housing affordability, gentrification, and displacement.

TRANSPORTATION

Transportation is one of the largest drivers of GHG emissions in cities and in many U.S. cities is the single largest source of GHG emissions. Policy approaches available to cities to address this issue include investing in public transit and bicycle and pedestrian infrastructure, implementing congestion pricing, establishing low-emission zones, and scaling up electric vehicles.

Legal questions that arise around transportation relate to preemption by federal laws like the Clean Air Act and the Energy Policy and Conservation Act; interplay with state public utility laws and with owners of private property in scaling up electric vehicles; overlapping federal, state, county, and municipal authority over road tolling and regulation; the privacy implications of monitoring traffic or mileage; and local land use. Transportation policies also have significant impacts on equity, particularly on access to transit and on local air quality, that may require additional legal analysis.

ENERGY

The variety of policy options available to help cities scale up the generation of green energy and phase out fossil fuel sources of power raise legal questions distinct from those relating to reducing emissions from buildings. With the increasing electrification of transportation, and the prospect of electrifying industrial processes and manufacturing, an increase in clean energy may also lead to reductions in GHG emissions in multiple sectors.

Green energy policy tools include those at both the distributed scale—like rooftop solar panels, community solar programs, and microgrids—and the utility scale, including large purchases of renewable energy or renewable energy credits (RECs) through green tariffs, power purchase agreements, virtual power purchase agreements, and unbundled REC contracts, as well as facilitating the development of in-city utility generation.

Local governments can encounter significant legal hurdles as a result of the federal and state legal and regulatory regimes that govern energy distribution and use. These include difficulties identifying an appropriate clean energy procurement strategy that comports with state law. Local governments with municipal utilities may face different questions from utilities served by investor-owned utilities or rural cooperatives. In addition, local governments must navigate legal boundaries in negotiating franchise agreements with utilities and in choosing whether to make some aspects of local energy service public. Distributed energy generation is also subject to legal requirements relating to building codes, land use, rate-making, and state preemption of local policies.

WASTE

The disposal and treatment of waste compose a small but significant percentage of U.S. cities' GHG emissions—around 5 percent.[4] Meaningful policy tools exist to reduce GHG emissions from waste while reducing reliance on landfills and addressing environmental injustices. These include increased recycling and organic waste collection, waste management requirements for construction, bans on single-use plastic items, and regulating commercial waste haulers.

Legal questions that come up here include limits imposed by the Dormant Commerce Clause of the U.S. Constitution and overlapping state and local jurisdiction over waste processing and disposal facilities, including with respect to siting and operations.[5] In addition, contracts with waste haulers and processors can involve questions of contractual law with broader implications. Local product bans and mandatory composting programs can also give rise to evolving legal considerations, including preemption concerns. An important aspect of waste policy is the reduction of waste transport emissions, which accrue to a city's accounting of transportation emissions but nonetheless are intertwined with waste disposal and processing. Because of historical patterns of siting waste processing facilities and landfills in low-income communities and communities of color, efforts to reduce waste and waste GHG emissions can have significant equity impacts.

Greenhouse Gas Emissions not Included in These Policy Categories

It's important to note which GHGs—and therefore which policy categories—are not considered in this primer.

The *GPC* includes sectors for inventorying emissions from "industrial processes and product use" (primarily "releases from industrial processes that chemically or physically transform materials") and from "agriculture, forestry and other land use (AFOLU)." These two sectors are highly specialized, and accounting for them in an inventory is complicated. The prevalence of industrial and AFOLU activities also varies considerably from place to place. Moreover, some industrial emissions are captured in an accounting of building sector emissions. Therefore, many cities' GHG inventories and mitigation policy portfolios begin with the categories of buildings, transportation, energy, and waste.

The *GPC* does not include *consumption-based emissions*. Under a consumption-based accounting, GHG emissions attributable to the manufacture, transport, and disposal of goods purchased by a city resident are counted toward that city's consumption-based GHG inventory rather than toward the place of manufacture, transport, or disposal. Similarly, travel by a city's residents outside the city's boundaries are counted toward that city's consumption-based emissions. There are valid policy rationales for both the sector-based and consumption-based emissions approaches,[6] but few U.S. cities inventory their consumption-based emissions, and the policy options available to cities for reducing consumption-based emissions are less well developed.

USING THIS BOOK

This primer presents an exploration of relevant legal issues that might inhibit or facilitate policy adoption across a range of municipal carbon mitigation policy areas: equity, buildings, transportation, energy, and waste. The exploration of what might be understood all together as urban climate law seeks to make this emerging area more readily comprehensible to a range of readers and to demystify necessary legal analyses so that researchers and advocates can pursue, and law- and policy-makers can craft, informed, creative carbon mitigation policies that address local political and policy concerns and stay within legal bounds.

1

CROSS-CUTTING LEGAL CONCEPTS

The carbon mitigation policies of the buildings, transportation, energy, and waste sectors give rise to unique and specific legal considerations, but these issues largely flow from legal concepts that municipal and environmental attorneys encounter again and again. This chapter provides an overview of the fundamental legal principles at play in urban climate law.

THE FUNDAMENTALS OF MUNICIPAL LAW

The city is the creature of the state.

—*Trenton v. New Jersey*, 262 U.S. 182, 189–90 (1923)

Cities are "merely subdivisions of the state," meaning that they have only the authority delegated to them by the state in which they are located.[1] The authority granted to municipalities by states varies significantly, both between states and within them. For any carbon mitigation policy, questions may arise concerning the extent of the municipality's delegated authority or the possibility that state or federal law might preempt it.

Municipal Home Rule and Dillon's Rule

While grants of municipal authority vary substantially in their details, two basic concepts of municipal law provide a framework for assessing a municipality's authority:

- Municipal Home Rule: A municipality has been delegated a broad range of authorities under a state law. In some Home Rule jurisdictions, a city can adopt a charter that serves as its constitution and allows for significant latitude to self-govern.
- Dillon's Rule: A municipality has no authority beyond what is granted expressly by state law, plus some incidental authority necessary to effectuate the express delegation of authority.

Home Rule and Dillon's Rule exist along a spectrum. In some states, and with respect to some areas of the law, a municipality will have significant latitude to legislate as it sees fit, as long as state law does not preempt such action. This is the Home Rule end of the spectrum. In other states, and with respect to other areas of the law, a municipality may legislate only as explicitly authorized by applicable state law. This is the Dillon's Rule end of the spectrum. Some states follow Home Rule for some municipalities and Dillon's Rule for others. No matter the particular approach of a jurisdiction, the underlying principle remains the same: cities have no more authority than what is delegated by the state.

STATE PREEMPTION OF LOCAL LAWS

State law has primacy over local law, and state laws can preempt local laws. State law may preempt local law in two basic ways:

- Conflict preemption: A state law and a local law conflict directly such that the local law cannot stand.
- Field preemption: The state legislature indicates, expressly or implicitly, that by legislating in an area, it intends to "occupy" the entire field.

Preemption by state law can arise in any area of urban climate law, and the landscape is notably varied across states. In assessing the legality of any law or requirement to reduce greenhouse gas (GHG) emissions, one must not only look for a source of authority from state law but also confirm that no state law preempts the proposed local action.

Targeted State Preemption Laws

Historically, state preemption of local climate law and policy has been incidental, a function of preexisting state laws governing the environment, energy, natural resources, and land use that took on new relevance in light of local efforts to mitigate carbon emissions. More recently, however, some Republican-controlled state legislatures have sought to directly preempt local climate laws. This wave of preemption laws can be traced to efforts to thwart the building electrification or "natural gas ban" movement (described in chapter 3). Many of these preemption laws, including the first, in Arizona, avoid an express focus on natural gas but are clearly aimed at stopping cities from imposing restrictions on the use of natural gas. Other preemption laws, like those in Florida and Indiana, are more explicit.[2] Approximately twenty states now have laws in place preempting local building electrification requirements and, in some instances, other local decarbonization policies.

The full scope and effect of these anticlimate preemption provisions remains to be seen. Their broad language could preempt many of the goals, strategies, and tools described in this primer. However, some elements of the state preemption laws may themselves be subject to legal challenge.

FEES AND TAXES

The law differentiates between fees and taxes. In most jurisdictions, a *fee* cannot be designed to raise revenue, must be tied to the cost of providing a benefit, and must be voluntary, in that an individual can avoid paying it by choosing not to obtain the associated service or benefit.[3] A *tax*, on the other hand, doesn't require that the payer receive any specific benefit, but taxes must follow the "uniformity principle": taxpayers, within reasonably established classes, must be treated equally. States vary as to whether they allow cities to establish new taxes without prior state legislative authorization.

The distinction isn't rhetorical. Many courts hold assessed amounts that do not meet the legal definition of a fee to be taxes, which a given city may not have authority to levy or, if it does, may have to adhere to different legal requirements to levy. In the climate context, this distinction means that some cities have limited authority to implement policies that price carbon or carbon-emitting activities. Pricing carbon at the city level is still a new idea. Cities interested in carbon pricing but without the authority to tax must carefully tailor pricing to meet the definition of a fee or work with their respective states to pass enabling legislation.

PENALTIES

Municipalities generally have the authority to impose fines, but this authority must be delegated by the state—either expressly or implicitly—and it differs from place to place.

A few basic rules include the following:

- Penalties must be reasonable. What this means varies depending on jurisdiction and the nature of the offense. The Eighth Amendment of the U.S. Constitution, which establishes the basic standard, prohibits excessive fines, as do similar provisions in state constitutions and various statutes.
- Penalties generally cannot be designed to raise revenue.
- The criteria for civil and criminal penalties differ and vary by state. Courts apply as many as seven factors in determining whether a penalty should be deemed civil or criminal.
- State law can impose other limitations on municipalities, such as maximum penalty amounts or delegation of penalty authority only for certain ordinances.[4]

Cities setting enforcement mechanisms or penalty amounts for local climate laws must operate within these parameters.

ROAD TOLLS

States also vary in whether they permit local governments to collect tolls on local roads. In New York, municipalities cannot collect tolls without specific authorizing legislation from the state.[5] In Oregon, they can, though there are restrictions on how tolling revenues can be used.[6] And in Washington, localities can create "transportation benefit districts" that have the authority to toll as long as such tolls are approved by "a majority of the votes in the district voting on a proposition at a general or special election."[7] Wherever a toll would apply to a federal-aid highway, federal approval would generally be required.

As a result, some cities have an easier time than others in finding the requisite authority to implement congestion pricing and other tolling schemes meant to reduce driving and its associated GHG emissions.

REALLOCATION OF AUTHORITY

As a general matter, local officials—who have either been elected by constituents or appointed by an elected representative—cannot give away (or delegate) their authority to make decisions on behalf of the municipality unless specifically allowed by law. In New York, for example, any law that "abolishes, transfers or curtails any power of an elective officer" must be put to the voters in a referendum.[8] In many states, municipalities are prohibited from trading away their legislative authority by contract.[9] Cities must follow applicable state law to determine which policy decisions are appropriate for administrative agency rule-making or being informed by input by an appointed board and which must be retained by the executive or legislature.

Why does this matter? Questions regarding the allocation of authority could arise, for example, as a municipality considers establishing a new climate office or officer or giving a significant role in a GHG emission–reducing activity to a private contractor. Some new climate laws and programs require the appointment of an expert or citizen advisory committee. These committees can generally provide advice and input, but ultimate policy-making authority must rest with a city's elected or appointed officials.

FEDERAL LAW

Because municipalities are political subdivisions of states, they are subject to the same limitations of federal law that states are.

Federal Preemption

The Supremacy Clause of the U.S. Constitution establishes the primacy of federal law and regulation over state and local law.[10] Therefore, where a state or local law conflicts with existing federal law or where a federal statute "occupies the field," the federal law preempts the state or local law.

A variety of federal statutes could preempt city climate laws. The Clean Air Act and the Energy Policy and Conservation Act are particularly likely to preempt state and local climate laws. These laws are discussed at greater length in chapters 3 through 5.[11]

Constitutional Restrictions

The U.S. Constitution contains basic restrictions on how governmental actors, including state and local governments, may behave. Some of these are discussed elsewhere in this chapter, such as the Fourth Amendment restriction on unreasonable searches and seizures (related to privacy), the Fifth Amendment Takings Clause (related to land use), and the Eighth Amendment clause on excessive fines (related to penalties). Here, we note two others. First, the so-called Dormant Commerce Clause, which bars states and local governments from passing laws that impermissibly discriminate against out-of-state economic actors,[12] provides an important limitation. Climate-related laws and policies can generally be crafted to avoid impermissibly discriminating against interstate commerce. Still, some policies may be out of bounds. For example, a city government cannot mandate that local waste-processing businesses refuse out-of-state waste to avoid generating GHG emissions in processing that waste.[13] Second, the Equal Protection Clause of the Fourteenth Amendment, which prohibits state and local governments from denying any citizen "equal protection of the laws,"[14] may also be relevant. Litigants sometimes make equal-protection claims, as in cases regarding differential tolls for out-of-state residents and fleet pricing for rental vehicles.[15] Notably, claims in both instances have failed.

PRIVACY

A city's reduction of GHG emissions depends on measuring emissions, sharing information, tracking progress, and changing behavior. These activities all have repercussions for individual privacy and data security. Privacy law, such as it is, includes requirements under federal, state, and local law and can arise in civil and criminal contexts.

Though the most basic privacy protection is the Fourth Amendment protection from unreasonable searches and seizures,[16] privacy concerns have been raised with respect to tolling and congestion pricing programs (which use license plate–reading cameras or GPS data to track vehicle whereabouts),[17] smart meters (which track account-level energy use data more or less in real time),[18] and requirements for the disclosure of building energy use. The collection of credit card information or other personal data adds another layer of complexity. Because technology is constantly evolving, and with it the capability and capacity to capture and retain massive stores of personal information, privacy law tends to lag. For cities, data and privacy concerns raise questions about how to minimize potential exposure to legal liability.

LAND USE

Urban climate action is inseparable from city planning and land use decisions. Strategies like promoting mixed-use areas, transit-oriented development, and walkable and bikeable neighborhoods help reduce GHG emissions from the transportation sector and possibly from buildings. Zoning codes may be used to require, incentivize, or facilitate energy efficiency, renewable energy, more bicycle lanes, more green space, less vehicle parking, and much more.[19] Land use law incorporates elements of federal, state, and local law, including zoning and eminent domain. A few key legal issues related to land use are as follows:

- *Zoning* is the process, generally at the local level, of prescribing certain permitted and prohibited uses in different geographic areas of a city. While zoning ordinances are usually municipal laws, zoning is often expressly enabled by a state law that limits what a zoning ordinance may include. Zoning ordinances may also be preempted by conflicting state laws or be required to conform to a city's land use plan. Zoning has been held by the Supreme Court to be an appropriate exercise of a municipality's police powers.[20]
- *Eminent domain* refers to the power of a government to take private property for public use. Under the Takings Clause of the Fifth Amendment to the U.S. Constitution, a taking by eminent domain requires that just compensation be paid to the property owner.[21] States inherently hold eminent domain authority, which they may or may not delegate to municipalities. The Supreme Court has held eminent domain to be appropriate for achieving a range of public ends, including economic development.[22] The issue may arise for climate

cities promoting transit-oriented development, municipalizing a public utility, or establishing other climate-friendly amenities.

- *Regulatory takings* are a form of taking in which the government does not actually occupy or take title to physical property but rather, through lawmaking or other governmental restriction, deprives a property owner of "all economically beneficial use" of their land.[23] While based on the Takings Clause and reinforced in all fifty state constitutions,[24] a significant amount of federal and state case law helps define what is and is not a regulatory taking. A regulatory taking requires a government actor to pay "just compensation" for the property owner's loss,[25] an undesirable outcome.

PRIVATE LAW

Urban climate law also incorporates legal relationships entirely between private parties. A large portion of a city's GHG emissions each year is emitted by private actors, and only those actors can reduce them. Local law and policy can require, incentivize, or provide resources for these reductions, but they can't achieve them on their own. For example, the landlord–tenant relationship gives rise to the so-called split-incentive problem: landlords are in a position to make energy-saving capital improvements to a building, but tenants are positioned to reap most of the energy-related cost savings.[26] The two could align their incentives through leases or other contracts, but a city has limited mechanisms to force that to happen. Many other private relationships contribute to—or help mitigate—a city's GHG emissions; the relationships between local businesses and commercial waste haulers, for example, have implications for both transportation and waste-associated GHG emissions.

REGULATING NEW BUSINESS MODELS

City carbon reduction policy must sometimes develop rapidly in response to outside market forces, involving several areas of the law. One example is micromobility: the use of shared bikes, electric bikes (or e-bikes), and scooters. Micromobility companies entered new markets by leaving dozens or hundreds of shared bikes or scooters on city streets. But cities hadn't yet developed regulatory regimes for this business model and had to solve issues related to public rights-of-way; insurance, indemnity, and bonding; safety; parking; data sharing; customer service; device types; and permitting.[27] While bike- and scooter-share show some (perhaps limited) promise in reducing emissions by compelling mode shift and offering last-mile, low-carbon transportation options,[28] they have caused some growing pains for cities. Cities may need to be flexible and adapt to new legal frameworks in response to entrepreneurial entrées into city-level efforts to reduce GHG emissions.

CONCLUSION

A key early step in assessing the legal frameworks applicable to a local climate policy is issue spotting: identifying as many potential legal issues as possible with the policy and its likely effects. This chapter provides a road map for spotting obstacles that can arise with proposed local laws and policies to reduce GHG emissions in any sector and can serve as a starting point for practitioners and students assessing a policy area. The chapters that follow take a deeper dive into how these and other issues appear in specific areas of city climate action.

2

EQUITY

Climate policy and issues of justice and equity are interwoven at all levels of government but perhaps nowhere more so than locally—in the places where people live and build community. Equitable climate policy is not a one-size-fits-all proposition. Goals may include reducing local air pollution in communities historically overburdened by harmful pollutants like carbon monoxide, nitrogen oxides, sulfur oxides, and particulate matter; investing and increasing employment in disadvantaged communities; and expanding the input of disadvantaged communities on local climate policy, among other things.

Some U.S. cities have well-developed climate justice or equity plans—policy road maps that center the concerns and well-being of environmental justice communities, low-income residents, and others experiencing frontline climate impacts. Many others do not. But nearly all the approaches in this primer have the potential to provide new opportunities for environmental justice communities—or to increase the burdens on them. Indeed, a "climate city" can—and should—be one with reduced local pollution, increased economic opportunity, and protections for the communities most harmed by climate change.

This chapter highlights the legal implications of equitable climate policy in the context of a local government.

DEFINITIONS

The terms *equity* and *climate justice* do not have single agreed-upon meanings that fit every circumstance. In the international context, for example, these concepts may refer to the obligation of nations that have contributed most to climate change to reduce GHG emissions or compensate others for loss and damage. In U.S. cities, equitable policy often aims to address environmental injustices, economic inequality, segregation, and systemic racism.

In their climate policy-making, local governments often develop their own working definitions. For example, the City and County of Honolulu offer five types of equity for consideration: (1) *procedural equity* ("accessibility and inclusivity of decision-making processes by those most impacted"); (2) *distributional equity* ("benefits are distributed to prioritize those most in need"); (3) *structural equity* ("transparency and accountability are institutionalized and regulated"); (4) *intergenerational equity* ("decisions prioritize the health and well-being of future generations"); and (5) *cultural equity* ("the acknowledgment and undoing of racism with the concurrent construction of multicultural norms").[1]

Austin, Texas, offers a definition of climate equity that expressly incorporates racial equity, where "racial equity is the condition when race no longer predicts a person's quality-of-life outcomes in our community."[2] In this view, equitable climate policy requires grappling with a "history of racial segregation and EJ [environmental justice] issues in Austin" to redress the disproportionate harm of climate change to residents who identify as BIPOC (Black, Indigenous, and people of color). While the wording—and even the substance—of local governments' defined terms and articulated objectives for equitable climate policy may vary, common themes align them, particularly with respect to righting procedural and distributional wrongs. To understand a local government's or community's goals with respect to equitable policy, it is necessary also to understand the local context of that government or community.

A NOTE ON RACE AND EQUITABLE POLICY

One of the more contentious legal questions that confronts equitable local climate policy-making is whether and how race may be a factor in developing climate policy and determining who benefits from that policy. In the relevant parlance, policies may be *race neutral*, meaning that they do not take race into account, or *race conscious*, meaning that they do. Racial categorizations can play an important role in climate policy-making because of the disproportionate historical and ongoing harms to low-income communities and communities of color caused by climate change and climate solutions. The topic is interwoven throughout this chapter, starting with a discussion of the constitutional right to equal protection, which sets the legal guardrails around how race may play into governmental actions and decisions, and continuing through discussions of civil rights, spending commitments, and procurement policies.

SOURCES OF LEGAL RIGHTS UNDERPINNING CLIMATE JUSTICE

Equitable local climate policy draws from a variety of legal sources, including civil rights and housing laws, decades-old environmental laws, relatively recent climate laws and executive orders, and the U.S. Constitution. It spans federal, state, and local law and often makes novel use of a familiar statute or legal concept to advance equity-focused or environmental or climate justice goals (even though these terms are not synonymous, the law relating to one can often be used to inform an approach with respect to another). The next section explores constitutional protections, federal civil rights and environmental legislation emerging in the 1960s and 1970s, evolving federal action on climate and environmental justice, and efforts at the state and local levels, all of which can (with varying degrees of success) play into legal arguments relating to equitable local climate policy.

EQUAL PROTECTION UNDER THE U.S. CONSTITUTION

Under the Fourteenth Amendment to the U.S. Constitution—which contains the Equal Protection Clause—and the case law interpreting it, governmental classifications based on race are subject to "strict scrutiny," a standard of review that requires a showing of a "compelling governmental interest" and a governmental response "narrowly tailored" to respond to that interest. This requirement can make it challenging for cities to develop climate policies aimed at mitigating climate impacts on communities and residents of color.

City of Richmond v. J. A. Croson, a Supreme Court case relating to local procurement policies, is instructive. There, the Supreme Court held that a local procurement policy that required prime construction contractors to award at least 30 percent of the dollar amount of a prime contract to subcontractors that were "minority business enterprises," or MBEs owned by individuals from underrepresented groups did not pass strict scrutiny review.[3]

While the court allowed that evidence of discrimination in the local construction industry could satisfy the "compelling governmental interest" component of strict scrutiny analysis, it was not sufficient in this case, where the city pointed only to a general history of discrimination in the construction industry. Moreover, the court held that, even if discrimination could be shown with "the particularity required by the Fourteenth Amendment," Richmond's response was not "narrowly tailored" to remedying the discrimination: it was not geographically restricted, and there were no waivers for contractors unable to meet the 30 percent requirement. In the years since *Croson*, however, many local governments have been able to demonstrate that local discrimination has risen to the level of a "compelling governmental interest" and narrowly tailor their procurement policies in response.

While equal protection law may help shape equitable climate policy moving forward, equal protection claims by private plaintiffs have thus far been largely unsuccessful in redressing environmental injustices. The Supreme Court has held that disparate "impact alone is not determinative" in establishing a violation of the right to equal protection[4] and that plaintiffs must show "an invidious discriminatory purpose."[5] And so, in *Bean v. Southwestern Waste Management Corp.*, plaintiffs demonstrated a discriminatory pattern in landfill siting but were unable to convince a federal district court in Texas of discriminatory intent. In *R.I.S.E. v. Kay*, the U.S. Court of Appeals for the Fourth Circuit noted the "disproportionate impact [of landfill siting] on black residents" but found that the siting board did not act in an "unusual or suspicious way."[6] Because it is often difficult or impossible to prove intentional discrimination, where a law or governmental policy is facially race neutral, equal -protection claims have provided little relief to environmental justice plaintiffs.[7]

Federal Civil Rights Law

Much like equal protection, federal laws aimed at protecting civil rights have also historically denied environmental justice plaintiffs redress for harmful environmental impacts. For example, Title VI of the Civil Rights Act of 1964 prohibits discrimination "on the ground of race, color, or national origin" by any activity receiving federal funding,[8] but Title VI litigation has not yielded significant environmental justice wins. The Supreme Court held that Title VI contains an implied private right of action to enforce Section 601, which prohibits intentional discrimination, but that Section 602—under which federal agencies promulgate civil rights regulations—contains no such right of action.[9] The U.S. Court of Appeals for the Third Circuit further held that plaintiffs had no private right of action under Section 1983 (which allows individuals to sue the government for civil rights violations) to enforce Section 602 regulations.[10] In other words, private citizens can sue to enforce the federal government's underlying obligation not to discriminate against individuals (i.e., the rights protected by Section 601) but not to enforce federal agency compliance with the regulations that such agencies promulgate to ensure they act in nondiscriminatory ways (i.e., the duties of federal agencies set forth in Section 602).

Civil rights law may be relevant in three ways. First, though it has not been fruitful for environmental justice plaintiffs, new fact patterns may merit another look at this type of claim. Second, federal and state agencies may increasingly view state and local projects through a Title VI lens to avoid disparate impact discrimination or litigation alleging such discrimination. Third, federal agencies may promulgate regulations to further protect against disparate impact discrimination, and local governments, among others, could be found to have violated Title VI by violating these regulations.

The National Environmental Policy Act

The National Environmental Policy Act (NEPA) is a federal law requiring federal agencies to take a "hard look" at the environmental impact of certain actions, including local projects involving federal funding or approvals. It is the federal version of a legal requirement often called "environmental review." While NEPA is not aimed specifically at environmental or climate justice or at racial or other kinds of equity, its requirement that federal agencies consider the environmental impacts of major projects has been wielded to require examination of impacts in low-income communities and communities of color. For example, the U.S. Court of Appeals for the D.C. Circuit found that an analysis by the Federal Energy Regulatory Commission of the environmental justice impacts of liquefied natural gas pipeline and export terminals was "arbitrary and capricious" because the Commission limited its review to communities located within two miles of the facilities despite impacts being projected to extend much farther. (The Commission was required to conduct further review.) Similarly, a federal district court in Wisconsin held that a review by the Federal Highway Administration and the Wisconsin Department of Transportation of a highway expansion project was insufficient because it failed to study the cumulative impacts of highway expansion projects on neighboring communities.[11]

NEPA is largely procedural; where it applies, the relevant federal agency must study the environmental impacts of their actions, but the remedy for a NEPA challenge is merely for the agency to perform more review. However, a NEPA review may lead to changes in or the cancelation of a project in at least two ways. First, agency decision-makers may choose to alter the project based on NEPA findings (and advocacy or litigation pushing them to do so). Second, a project delay can push important milestones back such that the project is scaled back or canceled. Some state environmental review laws, while structurally similar to NEPA, have more substantive requirements to mitigate environmental harms.[12]

Developing Federal Law

Had this book gone to press as recently as 2020, it would have highlighted a lack of meaningful protections for climate and environmental justice under federal law, aside from a 1994 executive order by President Clinton that directed federal agencies to incorporate environmental justice into their missions and policies.[13] In the past few years, however, federal law has developed rapidly, though in ways that have not yet played out in communities. First, in early 2021, came President Biden's Executive Order 14008, "Tackling the Climate Crisis at Home and Abroad," which sets out the administration's "policy to deploy the full capacity of its agencies to combat the climate crisis . . . [and] deliver[] environmental justice," including providing for "substantive engagement by stakeholders, including State, local, and Tribal governments."[14] In particular, the order made a commitment referred to as "Justice40," which sets a goal for certain kinds of federal climate and energy spending to be invested in a way such that "40 percent of the overall benefits flow to disadvantaged communities."[15] In addition, Biden issued a memorandum on discriminatory housing practices, which directs the U.S. Department of Housing and Urban Development to review Trump-era rules that may have weakened the Department's statutory obligation to "affirmatively further fair housing."[16] These federal orders and directives come into play only with respect to federal actions, but they may be useful to cities in advancing equitable climate policy where federal funding, approvals, permits, facilities, or other discretionary measures are involved. This includes a wide array of projects located in and directly affecting cities. Federal action can also serve as a model for local policy.

The Biden administration has also increasingly used legal tools like the Civil Rights Act of 1964 to intervene in state and local decision-making about projects that would have significant and unjust environmental impacts. For example, in 2021 the U.S. Department of Transportation cited Title VI of the Civil Rights Act of 1964 when it paused a proposed highway expansion in Houston to review its impacts on communities of color.[17] Federal action flowed through to other policy-makers, such as the Wisconsin Department of Transportation, which paused a highway expansion project in Milwaukee to perform further environmental review. This action was not required by federal law or the federal government, but the state transportation secretary noted that the supplemental environmental review would "help us make certain that our efforts to ensure racial equity with this project are comprehensive and aligned with federal priorities."[18] Similarly, the U.S. Environmental Protection Agency intervened in the permitting application process for a metal-shredding facility in Chicago, asking for a delay in the permitting decision pending review of environmental justice implications for "a community overburdened by pollution."[19] The city quickly responded by delaying the permitting decision[20] and ultimately rejected the application.[21] In these ways, federal agencies have shown how existing executive authority can be directed to protect communities vulnerable to environmental and economic inequities.

Congress has also taken significant action. In two marquee laws, Congress has altered its approach to issues of environmental and climate justice through provisions meant to benefit "disadvantaged communities," "energy communities," and other low-income communities and communities of color. The Inflation Reduction Act of 2022, the largest U.S. investment in climate mitigation and clean energy technologies to date, gives extra incentives for investment in disadvantaged communities and offers higher tax credits and up-front cash for electric vehicles and home energy improvements to people living below certain income thresholds.[22] The Infrastructure Investment and Jobs Act of 2021 appropriates federal dollars for investments in clean water infrastructure, pollution remediation, and more.[23] While both acts are spending laws and do not grant affirmative rights to communities beyond enhanced access to the laws' appropriations, the way these laws direct federal dollars could have positive impacts for frontline communities.

State and Local Climate and Environmental Justice Laws

Increasingly, state and local laws are tackling equity alongside GHG emission reductions. Some of these laws have direct implications for local climate policy-making. For example, New Jersey's 2020 environmental justice law mandates, with some exceptions, permit denials for facilities determined through an environmental justice review to, "together with other environmental or public health stressors affecting [an] overburdened community, cause or contribute to adverse cumulative environmental or public health stressors in the overburdened community that are higher than those borne by other communities" in the state.[24] New Jersey is considered to have more than three hundred municipalities that contain "overburdened communities." The climate laws of New York State and New York City also contain protections for frontline communities. New York State's Climate Leadership and Community Protection Act requires that, where GHG emission offset projects are used, projects that offer localized benefits to "disadvantaged communities" be prioritized and that the cumulative impact of GHG emissions in disadvantaged communities as a result of measures to reach net zero be considered.[25] New York City's law requires that, in reducing emissions from city government operations, the city should use "methods to ensure equitable investment in environmental justice communities that preserve a minimum level of benefits for all communities and do not result in any localized increases in pollution."[26] The city's law also requires a study regarding a potential emissions trading program for buildings to consider strategies that do not result in increased local air pollution.[27]

More traditional state and local environmental review laws are also playing an increasing, if still modest, role in encouraging equitable siting decisions in ways that could have implications for city climate policy. For example, the U.S. Court of Appeals for the Fourth Circuit recently vacated a Virginia state permit for a new compressor station, citing both a state law requiring consideration of the "suitability of [a project] to the area in which it is located"[28] and the "potential for disproportionate health impacts on the predominantly African-American community."[29] While state and local environmental review laws are often modeled on NEPA, they can contain substantively different terms. For example, some mini-NEPAs, like those in California[30] and New York,[31] require at least some effort to mitigate adverse environmental impacts identified during review. Land use law can play a role here, as well; for instance, some cities, like New York and Seattle, require or otherwise conduct an analysis of racial justice or equity implications that could arise for certain land use determinations.[32]

Spending Commitments

State and local climate laws may also address equity in relationship to spending. New York State's Climate Leadership and Community Protection Act pledges that at least 35 percent of the "overall benefits of spending on clean energy and energy efficiency" will go to "disadvantaged communities."[33] Washington's 2021 climate law pledges at least 35 percent of proceeds from a cap-and-invest program to "provide direct and meaningful benefits to vulnerable populations within the boundaries of overburdened communities."[34] President Biden made a similar pledge for 40 percent of spending in Executive Order 14008.[35] The details of which projects will qualify and which communities will be served by these spending pledges are works in progress.

BARRIERS TO EQUITABLE POLICY IN EXISTING LOCAL LAW

Where Climate Policy Decisions Are Made

Inequitable climate policy is to some extent a procedural problem. Those with more money, time, and historical representation are overrepresented in the forums where policy is made. This disparity yields policies that perpetuate historical injustices and inequalities, an outcome exacerbated by the disproportionate impacts of climate change experienced by low-income communities and communities of color. Significant work remains to ensure that local policy decisions are made in a transparent, inclusive way with opportunities for public input. Ensuring that climate policymaking is more participatory could involve changes that affect not only multiple local agencies but also state law.

Decision-making around climate policy implementation also implicates distributional—or outcomes-oriented—equity. Often, larger and better-funded groups are more likely to have the funding and capacity to monitor policy implementation, whereas environmental justice and community groups may need to shift focus to new priorities once the policy design phase is complete. The voices of these latter groups can be lost in policy implementation decisions, meaning these groups are deprived of procedural equity *and* may suffer outcomes that are worse for the communities they serve. Climate policy is not necessarily equitable policy, and implementation decisions can exacerbate gentrification, displacement, and other inequitable outcomes.

Procurement and Contracting

Energy transition rhetoric often touts green jobs. But local procurement and contracting policies may disfavor efforts to hire members of frontline communities for these jobs. Some cities have policies favoring businesses owned by women or individuals from underrepresented groups (often called minority-owned business or minority- and women-owned business enterprises), but even these may not be sufficient. Relatedly, sound local climate policy-making often relies on the input and expertise of frontline community members, who can provide critical information about both community and individual needs and the potential consequences of proposed policies. Local procurement and contracting requirements may need to be reviewed to allow city agencies to formally partner with, *and pay*, individuals and community groups for their work in convening stakeholders or informing city policy. Existing provisions of state or local law may inhibit some cities' ability to do this. For instance, some jurisdictions will have "race-neutral" requirements, meaning that race may not be taken into account in policy decision-making. (We do not mean to suggest that race-neutral policies are universally problematic. Rather, we mean to note that they can inhibit antiracist or redistributive climate policy.) Legal tools that local governments may have at their disposal to make hiring and procurement practices equitable include using community workforce agreements and project labor agreements, using existing authority to ensure acceptable compensation and workplace standards, and enforcing workplace requirements.[36]

BARRIERS TO EQUITABLE POLICY IN STATE LAW

As discussed in chapter 1, state law defines the contours of local authority; as a consequence, it can preempt local efforts to advance equity. In some instances, a state law may directly preempt a policy designed to advance the dual goals of climate action and equity. In others, state legislators may preclude the community process that would lead to the development of such a policy. State law can also restrict the ability of municipalities to access needed funds.[37] State-driven funding limitations include *direct limits on available funds*, meaning that the state fails to provide needed funds to a local government or limits the amount of debt a municipality may take on, and *restrictions on the uses of funds*. Examples of the latter include a New York State constitutional provision that prohibits cities from spending local (but not state or federal) tax dollars to improve private property[38] and an Oregon law limiting the use of road tolls collected by cities to the construction and maintenance of "highways, roads, streets and roadside areas."[39] While projects like bike lanes and bus shelters are arguably permissible uses of this funding, other projects to advance an equitable transportation system—like larger-scale public transit improvements—may not be. A state may also *limit or prohibit local taxation authority*. This limitation can leave cities without the funds to undertake policies that would reduce GHG emissions and advance equity goals. Fees are different from taxes; they involve the exchange of money for a service and do not treat all payers within a class equally, as a tax would. If local governments are forced to structure climate expenditures as fees for services provided, these expenditures could cause inequitable investment in favor of those who are able to pay.

STATE PUBLIC UTILITY LAW

State public utility (or service) laws (addressed in more detail in chapter 5) can address a range of areas that have direct implications for equity, climate justice, and energy justice. (*Energy justice* refers to efforts to make the energy system more equitable and democratic.) Critically, the state public utility (or service) commission is usually the forum in which electricity rates are set, pursuant to state law that guarantees a return to the utility; thus, a city's proposed climate policies may affect energy costs for energy-burdened households without any mechanism to correct for rising energy bills.

Advocates for energy justice have championed an evolving vision for the municipalization of local investor-owned utilities, seeing publicly owned power as a potential way to move away from GHG-emitting sources of electricity and to ensure a transition to a clean power system that prioritizes the needs of frontline community members. Municipalization is governed by a complex combination of state and federal laws that dictate the large sums needed to buy out or otherwise make the existing utility whole. Depending on state law, quasi-municipalization tools like community choice aggregation may be more attainable for some communities. No major investor-owned electric utility has been municipalized in the last decade.

EQUITY AS APPLIED IN CITY CLIMATE POLICY

Throughout this primer, we discuss how city climate policy intersects with equity and climate justice. In this section, we summarize some of the significant nexuses.

Buildings

Building decarbonization has significant impacts on housing quality, housing costs, and neighborhood gentrification and displacement. These factors are complex and potentially contradictory. For example, building energy retrofits can improve housing stock for low-income residents but can also lead to gentrification, the process by which new, wealthier residents change the character of a neighborhood and displace existing residents. As another example, all-electric building requirements reduce indoor air pollution but can sometimes increase energy costs. It is therefore critical to ensure that frontline communities and others living in affordable or rent-stabilized housing have a say in policy-making. Building policy plays out in the context of state law, particularly with respect to building codes, public utility law, and the energy distribution system.

Transportation

Transportation has a significant impact on people's day-to-day lived experiences. An accessible, reliable public transit network has the potential to expand access to economic opportunity and improve public safety. Climate and equitable access to transit should therefore be considered in tandem. Moreover, while shifting from vehicles with internal combustion engines to electric vehicles is critical, many people are and will continue to be unable to afford a new vehicle or will choose not to purchase one. So, a holistic look at a city's or region's transportation system is needed, including public transit and active transportation options like bicycling and pedestrian paths. Finally, some cities are replacing portions of their municipal fleets with electric vehicles such as buses and garbage trucks, particularly in neighborhoods with high levels of local air pollution and those that house bus and truck depots where vehicles often idle. In many cases, low-income and BIPOC residents experience elevated exposure to local air pollutants, leading to disproportionately higher incidences of childhood asthma and other health conditions. The transition to electric municipal fleets can help lessen this burden.

Energy

The energy transition intersects with equity in at least three critical ways. First, shifting from fossil fuel–powered energy-generating facilities, which cause significant local air pollution, to clean resources like wind and solar has the potential to substantially improve local air quality in the communities near fossil fuel power plants, which are often home to low-income residents and residents of color. However, the transition can be rolled out inequitably, with older gas- and coal-fired facilities in frontline neighborhoods among the last to be shut down. Second, without governmental interventions, shifting to new energy sources—particularly electricity—can increase overall energy bills for already-burdened households, including those left connected to fossil fuel sources like natural gas. Local governments may not be able to counteract this effect on energy costs without state cooperation. Third, the energy transition will create new jobs, and questions around who will get those jobs—whether the local labor force or people from frontline and low-income communities and communities of color—are central to what is termed a "just transition" from an economy powered by fossil fuels to one powered by renewables.

Waste

Waste—in particular the siting of landfills, incinerators, and other waste-transfer or -processing facilities—has long been part of the discourse of environmental justice. Siting decisions for recycling and waste-to-energy facilities and truck depots will continue to play a role in equitable city climate policy. Some cities go further by reducing emissions (both GHGs and local air pollution) from waste-hauling trucks, offering equitable access to community composting programs, and ensuring that waste workers receive appropriate labor protections.[40]

Land Use

Land use law in the United States has facilitated, and in many instances effectively required, the racial segregation of neighborhoods across the country. In many places, single-family zoning was historically used as a tool to exclude people of color; even today, single-family zoning requirements often lead to racially segregated communities. Importantly, single-family-zoned neighborhoods generally require more driving and larger living spaces, both of which generate more GHG emissions. Changes to zoning requirements, which in some instances are being used to reduce GHG emissions, may remedy or perpetuate longstanding patterns of racial inequity. In recent years, some cities have implemented zoning changes that address both climate and equity. For example, Somerville, Massachusetts, updated its zoning code to include a number of climate-friendly measures, including allowing accessory dwelling units, effectively eliminating single-family zoning.[41] If property owners take advantage of this opportunity and include an accessory dwelling unit on their property, housing stock will increase, and ideally some of this new housing will be relatively more affordable.

EMERGING LAW AND POLICY

Significant work remains to fully delineate how the law and equitable local climate policy intersect, including how legal concepts like equal protection and newer state climate laws can inform and either facilitate or hinder equitable policy-making at the local level. As equity considerations overlay all other sectors that are the focus of efforts to reduce GHG emissions, this work will play out on several fronts. Moreover, equitable local policy inherently varies from place to place based on a community's needs, population, and applicable legal frameworks. The inclusion of stakeholder voices, particularly from low-income communities and communities of color, in policy decision-making is critical. As local governments and communities work together to better understand these questions, this body of law will evolve.

3

BUILDINGS

Buildings are almost always a city's first or second largest source of GHG emissions.[1] Buildings are also relatively easy for a local government to regulate, since they are stationary—wholly within a city's limits—and part of the built environment that traditionally falls within a local government's authority. And controls on building emissions reduce not only GHGs but also local air pollutants, making buildings healthier places to spend time and potentially upgrading the city's building stock, making the city also a more equitable community in which to live and work.

Three strategies underscore building decarbonization (figure 3.1).[2] Step 1, maximizing energy efficiency, reduces the energy consumed by a building and its occupants and, with it, the emissions associated with that energy use. Step 2, electrifying all building components and appliances, paves the way for a 100 percent carbon-free building once step 3, powering the electricity grid with all clean or renewable sources, is complete.[3] These "steps" can and should be undertaken concurrently. While not addressed in depth in this volume, the decarbonization of building materials (or "embodied carbon") and the deployment

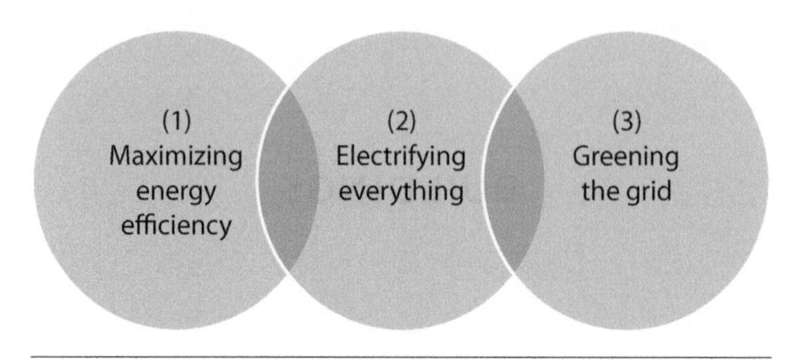

FIGURE 3.1 Primary strategies of building decarbonization.

of technologies that allow for energy use to be shifted to times of day with lower demand (called "demand-response" strategies) can be considered fourth and fifth aspects of building decarbonization (figure 3.2).

This chapter examines the legal issues associated with local government policy options addressing steps 1 and 2: maximizing energy efficiency and electrification. Chapter 5 addresses step 3: powering the electricity grid with clean or renewable sources.

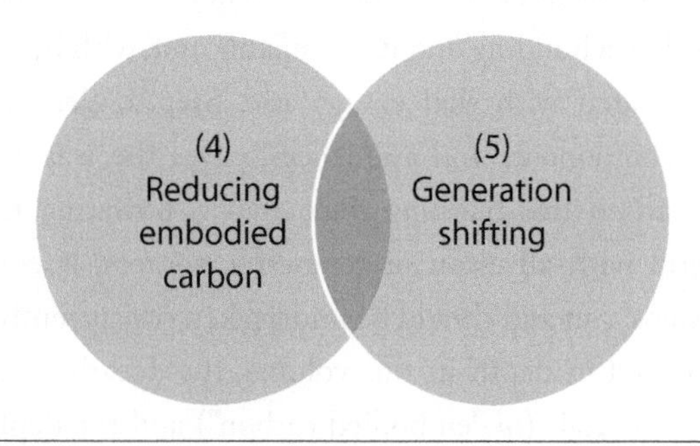

FIGURE 3.2 Secondary strategies of building decarbonization.

NEW AND EXISTING BUILDINGS

Building decarbonization strategies can be divided into two categories: those for *new* buildings and those for *existing* buildings. The construction of new buildings offers an opportunity to build in the most up-to-date, climate-friendly building features right from the start, and certain legal tools, such as building codes, offer a legal pathway to require just that. Existing buildings, in contrast, require retrofits that can be both costly and disruptive, making regulation more difficult. But buildings last a long time, and many existing buildings will continue to exist for decades—even centuries—to come. So decarbonization policies for buildings must address new *and* existing buildings.

WHERE DOES LOCAL AUTHORITY TO REGULATE BUILDINGS COME FROM?

The authority of cities to regulate buildings is relatively clear in most states. Building regulation is generally considered an element of *police power*, the broad authority left to the states by the Tenth Amendment of the U.S. Constitution to regulate with respect to public health and welfare, which many states delegate to municipalities through municipal home rule provisions, a specific grant of police power authority, or both.

Preemption by state law can significantly impede a municipality's ability to exercise police power, especially when the state retains all authority over building construction codes. Constitutional protections like equal protection, procedural due process, and substantive due process impose requirements of rationality, procedural compliance, and nondiscrimination. Building-related requirements are unlikely to constitute takings under the Fifth

Amendment of the U.S. Constitution, which requires property owners to be compensated for their losses, because, while green building requirements may impose costs on developers, they rarely if ever prevent the economic development of property.

SOURCES OF BUILDING EMISSIONS

Building emissions are generated almost entirely by the appliances within a building such as furnaces; heating, ventilation, and air conditioning (HVAC) systems; water heaters; washing machines and dryers; lighting; and computers and servers. Emissions generated on-site—as through the combustion of natural gas or other fossil fuel—are considered "scope 1" emissions, whereas emissions attributable to electricity and other forms of energy delivered to the building are referred to as "scope 2" emissions.[4] Both scope 1 and scope 2 emissions are counted in city GHG inventories. (Some of the policy tools addressing scope 2 emissions are covered in chapter 5, which discusses the decarbonization of energy systems.) Building decarbonization policies can address reductions in energy use or emissions at the level of individual appliances or the whole building.

Appliance Regulation

The federal government retains regulatory authority over the energy efficiency of many appliances used in buildings. So, while local governments may have authority to regulate *building* energy use, they are less able to regulate the *appliances* that directly contribute to a building's energy use. This chapter discusses the preemption of the state and local appliance standards of the Energy

Policy and Conservation Act (EPCA) throughout;[5] in their decision-making, policy-makers must be aware of the interplay between their building regulations and the federal regulations of the appliances within their buildings. Rules that regulate the non-energy attributes of appliances—such as standards relating to emissions—may be lawful for local governments, subject to questions of state and possibly federal law preemption.

BUILDING CODES FOR NEW AND RENOVATED BUILDINGS

Throughout this text, *building code* refers to a range of requirements that set standards for building construction or major renovations; for example, a building or construction code, an energy or energy conservation code, or a plumbing code, each of which may apply to residential, commercial, or some other subset of buildings. Often, a building code will incorporate other codes, such as an energy code, by reference. Existing buildings are required to comply with some elements of an applicable building code, but, because building codes largely regulate construction, they are treated here primarily as a tool for reducing GHG emissions from new buildings.

Statewide codes are often based on two model codes, the International Energy Conservation Code and the ANSI/ASHRAE/IES Standard 90.1, both of which are updated periodically. The Energy Policy Act of 1992 requires states to review their codes to determine whether revisions to the model codes "would improve energy efficiency" in the state's code.[6] State codes generally reflect the model codes but may be outdated or in amended form. In states where code authority is not delegated to local governments, statewide codes apply in all municipalities.

The building code is the key tool used to require, incentivize, and catalyze the development of new and renovation of existing buildings to be more energy efficient and produce fewer emissions. Decarbonization measures enacted through local building codes include the following:

- Solar readiness: requiring new buildings to be wired to accommodate solar panels or a solar hot water system
- Electric readiness: requiring new buildings to be wired to accommodate electric appliances, even if gas appliances are installed at the time of construction
- All electric or electric incentivized: requiring newly constructed buildings to be powered only by electricity, or offering incentives or compliance pathways to new buildings built without fossil fuels (or with fully electric or geothermal appliances and building systems)
- Other: more stringent requirements for the building envelope, water heating, lighting, plug load, equipment, or use of renewable energy

Building code authority is assumed by the states (as an element of police power), and only some delegate it to local governments. Local governments without building code authority are less able to amend construction standards in favor of improved energy conservation or to require the use of less carbon-intensive fuel sources. They may also face heightened preemption scrutiny with respect to construction requirements that they enact outside the building code framework.

This greater scrutiny occurs because statewide preemption not only implicates requirements codified *in* a building code but also might block the local adoption of construction-related requirements elsewhere, whether as a stand-alone local law or as part of another code. For example, a restriction on natural gas connections for new buildings in Brookline, Massachusetts, was disapproved by the Massachusetts attorney general's office, in part because the attorney general concluded that the restriction was preempted by the statewide building code.[7] Other building requirements, however, may not be preempted by a statewide code. For example, in Michigan, a court held that requirements relating to the maintenance of existing buildings were distinct from those covered in the statewide building code and therefore were not preempted.[8]

Even where cities *do* have building code authority, processes and substantive requirements to obtain state approval for amendments to building and energy codes vary by state. In some places, a local government must demonstrate a need for an amendment particular to the locality. In Pennsylvania, for example, a local building code amendment must be justified by "clear and convincing local climatic, geologic, topographic or public health or safety circumstances or conditions."[9] In New York, most municipalities must demonstrate "special conditions prevailing within" the locality to enact an amendment to the local building code (though these same New York municipalities have more authority to amend the building *energy* code).[10]

ZERO-EMISSION AND NET-ZERO BUILDINGS

Zero-emission and "net-zero" buildings are becoming the gold standard for new climate-friendly construction. These buildings have very low emissions in real terms and often zero emissions in net terms via the use of renewable energy credits. A building code is where a zero-emission or net-zero requirement for a building should be made, though for the most part state and local governments have not yet begun doing so. In 2021, Massachusetts enacted climate legislation containing a requirement that the state develop a net-zero *stretch code*: a code more stringent than the statewide base building code that municipalities may adopt if they wish.[11] California has also adopted building code requirements that move the state's code toward a net-zero standard, notably shifting to the use of electric building appliances like heat pumps and water heaters.[12] California's new code provisions follow the lead of many local codes around the state.

FEDERAL PREEMPTION OF STATE AND LOCAL STANDARDS FOR APPLIANCE EFFICIENCY

EPCA sets energy and water efficiency standards for "covered products,"[13] including many residential, commercial, and industrial appliances like furnaces, HVAC systems, and water heaters, as well as smaller appliances such as dishwashers, washing machines, and dryers.[14] Appliance energy use and building emissions are closely correlated; essentially all of a building's energy use is attributable to its appliances. Decarbonizing buildings therefore requires transitioning from fossil fuel–powered appliances.

EPCA preemption restricts state and local governments from regulating the energy efficiency or energy use of many appliances. Section 6297(b) of EPCA states that "for any covered product, no State [or local] regulation, or revision thereof, concerning the energy efficiency, energy use, or water use of the covered product shall be effective," unless one of a list of exceptions is satisfied.[15] A few of these exceptions are limited in scope to state and local laws enacted before 1987 or to specific local requirements enacted with respect to identified appliances,[16] for example, or to state laws granted a waiver by the U.S. Department of Energy.[17] For a local standard to be preempted by EPCA, the standard must apply to both a "covered product" and "concern[] the energy efficiency [or] energy use" of such product. In April 2023, the U.S. Court of Appeals for the Ninth Circuit took a broad view of EPCA preemption by invalidating Berkeley, California's restrictions on natural gas piping in newly-constructed buildings, suggesting that even local laws that do not set express or facial energy efficiency or energy use standards may be preempted by EPCA.[18]

Three notable exceptions to EPCA preemption are available to some local governments. First, local governments may set procurement requirements above federal standards.[19] Second, local governments may set efficiency standards for products not covered by EPCA standards, such as computers and computer monitors.[20] (This exception remains subject to state law, and a city's standards may be preempted by state-level standards. At least eleven states have some appliance efficiency standards on the books.[21]) Third, local governments with building code authority have some limited authority to set appliance standards that would otherwise be preempted by EPCA as long as those standards are part of a building code that satisfies seven statutorily defined conditions.[22]

THE BUILDING CODE EXCEPTION TO EPCA PREEMPTION

A building code may include standards that relate to the energy efficiency or energy use of EPCA-covered appliances if the code meets the following conditions: the code (1) "permits a builder to . . . select[] items whose combined energy efficiency" meets an overall building energy target; (2) does not specifically require any EPCA-covered appliance to exceed federal standards; (3) offers options for compliance, including one or more options that use appliances that meet but do not exceed federal standards; (4) bases any baseline building design used by the code on a building with covered products that do not exceed federal standards; (5) offers at least one "optional combination[] of items" that does not exceed federal standards for any covered appliance; (6) frames any energy target as a total for the building; and (7) uses EPCA-specified test procedures to determine the energy consumption of covered products.[23] In essence, these conditions mean that local building codes *can* set building-wide energy use or conservation standards that account for appliances exceeding federal standards but that they cannot effectively *require* the use of those particular appliances.

Two cases demonstrate what can and cannot be done. In *Air Conditioning, Heating & Refrigeration Institute v. City of Albuquerque*,[24] a federal district court examined the 2007 building code of Albuquerque, New Mexico, which offered performance and prescriptive options for compliance. The performance-based options effectively required the installation of appliances that exceeded federal appliance standards. The prescriptive pathways offered alternative options for energy conservation that a builder could undertake if they chose to use appliances that met but did not exceed federal standards. Those alternative options generally required developers to include additional products, incurring additional costs. Although the code gave builders options for compliance, the court held that the code effectively imposed "a penalty [in the form of increased costs] . . . for selecting products that meet, but do not exceed, federal energy standards," thus "effectively requir[ing] the installation of products that exceed" EPCA standards. The court therefore held Albuquerque's code preempted.[25]

A few years later, the U.S. Court of Appeals for the Ninth Circuit decided *Building Industry Ass'n of Washington v. Washington*

State Building Code Council,[26] which considered Washington's 2009 state-level building code. The code allowed for three compliance pathways to achieve an overall reduction of 15 percent in building energy use. Two pathways would not by themselves meet the target, and builders would therefore be required to earn one "credit" for other energy conservation measures. Credits could be earned, among other ways, by installing appliances that exceeded EPCA standards. The court held that the code was not preempted by EPCA, distinguishing Washington's code from Albuquerque's by noting that the Washington code did "not create any penalty or legal compulsion to use higher-efficiency products."[27]

LOCAL GOVERNMENTS WITHOUT BUILDING CODE AUTHORITY

In states where building codes are set at the state level and provide no opportunity for local amendment, municipalities can find themselves stymied in their efforts to decarbonize new buildings. A couple options are available, though.

First, these local governments can engage in advocacy to state-level law- and policy-makers, pressing their state's elected officials and administrative authorities to strengthen building energy code requirements for the whole state. For example, a municipality could push for the adoption of a statewide stretch code. Massachusetts has a stretch code that emphasizes improvements in energy performance.[28] Vermont has one that addresses efficiency improvements and solar- and electric vehicle–ready requirements.[29] New York has a stretch code that aims to advance building energy standards by offering a vetted local code option that is "about one cycle ahead of" the state's base code.[30] In Massachusetts, adoption of the stretch code is required for participation in the state's "Green Communities" program, which provides resources to reduce energy use in municipal buildings.[31]

Another option is a *home rule petition* to the state legislature. In filing a home rule petition, a local government effectively asks the state legislature to pass a state law specific to the municipality. For example, the state law could allow the petitioning local government to adopt more stringent construction or energy requirements than those of the state code. The home rule process can be a highly political one, and some cities will find that the state legislature as a whole does not share the progressive views of their own representatives, but it is an available legal tool for cities looking to set local requirements that differ from statewide standards.

THE ELECTRIFICATION OF NEW BUILDINGS

An all-electric construction policy—sometimes called a "natural gas ban"—is a measure or set of measures aimed at prohibiting or strongly disincentivizing natural gas connections to newly constructed or significantly renovated buildings.

Sources of Local Authority

Cities have relied on three main sources of legal authority for all-electric requirements:

- First, some municipalities have leveraged their local building code authority to enact all-electric construction requirements, including more than fifty in California alone.[32] These requirements vary widely, with varying exceptions and incentive structures, but have in many cases been approved by the California Energy Commission as permissible local amendments to the building energy code.[33] And Seattle has enacted all-electric construction code requirements for commercial buildings, effectively banning natural gas from new buildings.[34] New York City has adopted a different formulation, requiring through the local building code that newly constructed buildings emit no more than a set amount of GHG pollution per unit of energy used.[35]

- Second, some municipalities rely on their municipal home rule authority or their police powers.[36] This approach has encountered federal and state law setbacks. Berkeley, California's prohibition on gas infrastructure in new buildings was invalidated by the U.S. Court of Appeals for the Ninth Circuit, which in April 2023 held Berkeley's ordinance preempted by EPCA (as of this writing, Berkeley was reportedly considering petitioning for *en banc* review of the decision in the Ninth Circuit but had not formally filed a petition).[37] In Massachusetts, the town of Brookline's ban was disapproved by the state's attorney general.[38] Restrictions on natural gas that rely on home rule authority or the police power may yet be found lawful in other states and outside of the Ninth Circuit, but courts outside of these jurisdictions have not yet considered these kinds of local policies.
- A third potential source of authority is local zoning authority, through which a city may offer incentives for fossil fuel–free construction. While this approach was unsuccessful in Brookline, Massachusetts, where the attorney general's office disallowed two local zoning incentives aimed at catalyzing natural gas–free construction, other local jurisdictions may have more legal authority to adopt such incentives.[39]

State Law Preemption

Local all-electric construction requirements may confront state preemption problems. For one thing, a growing number of states are enacting laws expressly preempting local natural gas restrictions for buildings. Echoing the response to past areas of local environmental leadership—think plastic bag and fracking bans—more than twenty states (as of 2023) now prohibit any local requirement that restricts natural gas connections to new buildings.[40] In states with statewide building codes, local governments are preempted from amending their local codes to require or incentivize all-electric construction. Public utility law—generally set at the state level—can also give rise to questions of preemption. For instance, some argue that the statewide regulation of utilities is intended to occupy the entire field of utility regulation, thereby preempting (through field preemption) local restrictions on natural gas connections in buildings. Others argue that state utility requirements may conflict with local prohibitions on natural gas. New York's public utility law states that natural gas is "necessary for the preservation of the health and general welfare and is in the public interest;"[41] this so-called "obligation to serve" language has not yet been held to preempt local natural gas restrictions in New York or in any other state but has given local policy-makers pause that it could.[42] This is not to say that state public utility law *would* preempt local restrictions on natural gas hookups based on these arguments or that it would do so in every state. Many argue that a building's connection to natural gas is not in the same field as the regulatory regime of a state public utility aimed at the sale and distribution of natural gas; therefore, the state's energy laws and regulations should not preempt local restrictions.

Other Legal Considerations

FRANCHISE AGREEMENTS

Franchise agreements between municipalities and natural gas utilities may have a role to play in the electrification of new buildings. Franchise agreements give utilities access to the public right-of-way to lay pipes, wires, and other infrastructure, often in exchange for a fee. Some argue that particular franchise agreements grant an implied monopoly to the utility, a line of reasoning that could be used to oppose policies that would inhibit growth in the utility's customer base. In some instances, franchise agreements can play a more productive role, particularly when an electric utility's franchise is soon up for renewal or has other opportunities for negotiation. As discussed in chapter 5, the terms of franchise agreements can be used to incentivize or contract for a shift away from natural gas.

ENVIRONMENTAL REVIEW STATUTES

A small number of local natural gas restrictions have attracted litigation on the basis that the municipality failed to complete the environmental review required by state law. Two homebuilders in California sued the municipalities of Windsor[43] and Santa Rosa,[44] contending that the local governments did not sufficiently take the environmental impacts of their all-electric code provisions into account. Specifically, the homebuilders argued that increased electrification could contribute to wildfire risk, a factor the homebuilders asserted was not properly considered. In response to the suit against it, the town of Windsor opted to drop its all-electric code requirements rather than litigate. A state court allowed Santa Rosa's code to stand.[45]

Case Study: Berkeley, California

Berkeley was the first city in the country to "ban" natural gas.[46] Like Brookline, Massachusetts, Berkeley used its police powers to prohibit natural gas connections to new buildings. (It later supplemented its police powers–based ban with an all-electric building code revision, as have many other local governments in California.) Berkeley was sued by the California Restaurant Association (CRA), which argued state and federal preemption of the Berkeley ban.[47] The CRA's state law claims asserted that Berkeley should have followed the state building code amendment process and other state laws in enacting its natural gas restrictions; its federal law claims asserted that EPCA preempted the Berkeley gas ban because the ban effectively prohibits the use of natural gas appliances and therefore "concern[s] the . . . energy use" of covered building appliances. Though a federal district court dismissed the CRA's complaint, in 2023 the U.S. Court of Appeals for the Ninth Circuit invalidated Berkeley's natural gas ban. The panel of judges held Berkeley's ordinance preempted by EPCA because it interferes with "the end-user's ability to use installed covered products at their intended final destinations."[48] The Ninth Circuit holding represents a significant expansion of EPCA preemption. The ruling applies only in the Ninth Circuit, and at the time of this writing a petition by Berkeley for review of the decision en banc (i.e., by a larger set of Ninth Circuit judges) was pending.

LAND USE AND BUILDING EMISSIONS

Many cities use their zoning codes to encourage sustainable building practices. Land use rules can—in some jurisdictions—

require or incentivize that buildings be constructed to a low-energy or low-GHG standard. In many jurisdictions, land use controls can place limits on building size or encourage smaller living spaces, reducing the per capita energy use of a building. There are many examples of zoning provisions that can reduce building energy use or GHG emissions. For instance, Watertown, Massachusetts, requires buildings of more than ten thousand square feet or with ten or more residential units to include a solar energy system.[49] Building setback and appearance requirements, among others, can allow for exterior building insulation or outdoor appliances like heat pumps or clarify that solar panels or other renewable energy technology is permissible. Cambridge, Massachusetts, for example, provides adjustments to building setback requirements to facilitate the installation of exterior insulation.[50] Zoning can require that new buildings achieve (or qualify for) certification from Leadership in Energy and Environmental Design (LEED) or another third-party building certification program. This is the case in Miami, where buildings of more than fifty thousand square feet must achieve LEED Silver certification or an equivalent level of performance.[51]

Zoning can also offer incentives—such as a density or floor area ratio bonus, a fee waiver, or a special permit—in exchange for various sustainability efforts. For example, in Minneapolis, downtown developers of buildings that are 35 percent more efficient than required by the Minnesota code are offered floor area ratio premiums;[52] in Miami Beach, the city refunds a portion of a development fee based on the building's level of LEED certification.[53] Zoning provisions that address building construction—widely understood to be in the domain of the building code—can invite preemption questions. Zoning incentives aimed at construction or energy may avoid preemption by a state building code more easily than by a zoning mandate.

Cities can also use land use authority to reduce building sizes or increase density, thereby reducing per capita energy use. For example, revising a local code to allow for accessory dwelling units or "tiny houses" means that some residents will live in very small homes that require little energy use per person. These code changes can also address housing scarcity or unaffordability and contribute to denser, possibly more walkable neighborhoods. The city of Spur, Texas, allows the construction of tiny houses by right in most areas of the city,[54] while cities like Ann Arbor, Michigan,[55] and Somerville, Massachusetts,[56] allow for accessory dwelling units. A handful of U.S. cities have eliminated single-family zoning entirely. Zoning codes can also set maximum building sizes; traditionally, local governments have set such restrictions to control for neighborhood uses and design, but they can also be aimed at limiting building energy needs. All these approaches fit comfortably within traditional notions of municipal zoning authority.

LEED AND OTHER THIRD-PARTY BUILDING STANDARDS

Some cities mandate or incentivize construction that meets the standards of third-party sustainability certification programs such as LEED (from the U.S. Green Building Council) and Passive House. Such requirements are often codified in local zoning and building codes. Although many avoid legal scrutiny, a handful of legal considerations can arise.

The wholesale adoption of LEED or another third-party sustainability rating system can give rise to questions under the nondelegation doctrine about whether public legislative authority has been improperly delegated. The nondelegation doctrine "limits a legislative body's delegation of its functions to other branches of government or to private entities."[57] Two types of nondelegation issues can arise when a city adopts a third-party building certification system. First, basing approval of a development's green attributes entirely on a third-party standard, without independent assessment, could raise red flags. Second, tying local requirements to evolving third-party standards could leave a city in a position in which changes to its building code, zoning code, or other building requirements could be made in the absence of legislative or regulatory process.

Further, third-party rating systems may not prioritize GHG emissions or energy use. The most widespread building rating program, LEED, offers points across nine categories, and a minimum number of points must be earned to achieve various levels of certification.[58] Only one category—energy and atmosphere—directly prioritizes reductions in GHG emissions. Others, such as water efficiency, contribute to a reduction in energy use but do not prioritize it. And some categories, such as innovation in design, may not address building energy use or GHG emissions at all. (Other ratings systems, such as Energy Star, do prioritize reductions in building energy use.[59])

A few strategies can help mitigate legal concerns. First, cities can write the requirements of a third-party standard into the applicable code, thereby avoiding a situation in which amendments to third-party standards change local building standards without any regulatory process. Second, cities can require that projects be certifiable under a third-party rating system, rather than fully certified.[60] Structuring requirements this way not only avoids any delegation of permitting decisions to a private entity but may also lessen compliance and administrative costs. Third, local governments can require LEED certification for only certain types of buildings, such as affordable housing and other large projects, as in Boston;[61] municipally owned and operated buildings, as in Houston;[62] or municipally owned, operated, or funded buildings, as in Washington, DC.[63] Finally, a local government can develop its own building sustainability rating system, as Seattle did with its Evergreen Sustainable Development Standard[64] and Chicago did with its Chicago Sustainable Development Policy.[65] This approach is resource intensive, but it retains legislative and regulatory authority within the local government and can prioritize preferred energy and GHG-related building attributes.

EXISTING BUILDINGS

Policies to reduce GHG emissions from existing buildings differ from those for new buildings. As a technical and cost matter, requirements for new buildings are often simpler to comply with, as they are considered early in the design and construction processes. As a legal matter, existing building regulations can often sidestep concerns about state building code preemption. But energy and emissions requirements for existing buildings are novel and evolving, and many potential legal questions have not yet been settled.

BENCHMARKING, AUDITING, AND DISCLOSURE REQUIREMENTS

Reducing GHG emissions from existing buildings often begins with benchmarking and energy disclosure. These requirements do not require buildings to reduce emissions or energy use; they simply help a city understand the energy and emissions profiles of local buildings and chart a path to emission reductions. You manage what you measure. These requirements also help building owners and managers assess their energy use as compared to other similar buildings and identify opportunities for improvements. In some instances, these requirements can provide information to property buyers and renters, potentially shaping the market in favor of properties that use less energy.

Benchmarking, auditing, and disclosure requirements differ. A benchmarking law requires buildings, often of a certain type or use category or above a certain size, to provide energy and emissions data for comparison to other buildings of similar type, use, and size.[66] Often, the data are entered into the U.S. Environmental Protection Agency's Energy Star Portfolio Manager online portal.[67] Benchmarking information can be used in the near term to identify opportunities for building owners to reduce GHG emissions or energy use.[68] This information can later inform regulatory requirements, such as building performance standards (discussed later in the chapter). In New York, city officials now use benchmarking data from the city's largest buildings to assign letter grades (i.e., A, B, C, D, or F) for building performance; these are posted near building entrances.[69] Chicago has a similar system that specifies a star rating for building energy performance.[70] Energy audits entail a more intensive look at a building's energy use, giving owners information about their use and potentially offering recommendations for cost-saving improvements.[71]

Disclosure policies use a variety of strategies beyond benchmarking and are more likely to apply to smaller buildings as well as larger ones. Some disclosure policies require energy-related information to be provided to potential property buyers, as under Minneapolis's time-of-sale energy disclosure requirement[72] and Portland, Oregon's residential energy performance rating disclosure requirements.[73] This information may give purchasers the opportunity to negotiate a lower price for a poor-performing property or choose among properties based on their energy use.[74] Requiring this information to be provided to potential renters—or simply requiring that it be publicly available—can also help shape the rental market, which is undermined by an imbalance in information between landlord and tenant.

There are two key legal considerations relating to benchmarking, auditing, and disclosure policies: privacy and preemption.

Privacy and Data Security

Some have raised concerns about privacy interests relating to building or unit energy use, particularly when the data can be used to identify an individual home or residential apartment. In one case, Naperville, Illinois, enacted a requirement that buildings contain smart meters that collected energy use data every fifteen minutes. The U.S. Court of Appeals for the Seventh Circuit held that the data collected were "searches" for purposes of the Fourth Amendment but that the city of Naperville's "significant government interests" (i.e., "the modernization of the electrical grid") in using such data made the search reasonable.[75]

The use of smart meters and other smart grid technologies can also give rise to cybersecurity concerns and related legal risks. Whether the municipality, a utility, or some other third party holds the data, the risk of breach or theft exists.[76] As a result, smart metering laws must comply with federal data privacy and protection laws, including the Electronic Communications Privacy Act[77] and the Computer Fraud and Abuse Act,[78] and state laws, such as the California Consumer Privacy Act.[79] Moreover, even if all applicable laws are followed, a city could still be sued in the event of a breach or other misuse of data.

The politics of smart metering can be tricky, but there are tools to address citizens' privacy concerns. For instance, a local government can limit public access to such data under applicable freedom-of-information laws or craft an opt-in or opt-out policy that requires building owners to pay the excess cost of collecting energy use information from buildings without smart meters.

Collecting Data from Utility Companies

One simple way to manage the collection of building energy data is to require that the local electric or gas utility provide it directly to the city, town, or county's building department. However, doing so may be legally infeasible. Public utilities are largely regulated at the state level, so requiring that a utility company turn over data may not be an option available to local governments. Still, a local utility company may be a good partner in developing this sort of policy, helping to streamline the administrative burden of compliance by individual customers. Cities with municipally owned utilities may have more latitude to direct their utilities to provide them with data.

EXISTING BUILDING PERFORMANCE STANDARDS

Building performance standards do exactly what their name suggests: require that a building meet a set metric of performance, such as a cap on energy use or GHG emissions, while allowing building owners discretion to determine *how* to meet the metric.[80] Performance standards can be contrasted with prescriptive requirements, which require building owners to undertake expressly specified measures (such as prescribed building upgrades, retrocommissioning, or building tune-ups) and decrease energy use or GHG emissions but do not necessarily set numerical requirements with respect to the results. So far, there are only a handful of building performance standards in the United States and even fewer for which compliance periods have begun. Many of these standards apply to larger buildings of twenty-five thousand or fifty thousand square feet or more.[81] Other cities have policies that can be considered building performance standards but are more limited in scope or have prescriptive options for compliance. For example, Reno, Nevada, requires buildings of more than thirty thousand square feet to meet one of four energy target options (including an energy-use intensity metric), but it offers several alternative compliance pathways.[82] Similarly, Boulder, Colorado, requires residential rental buildings to meet energy efficiency standards for the building owner to get a rental license, but an alternative prescriptive compliance pathway is available.[83] These hybrid policies allow local governments to experiment with performance standards while giving building owners flexibility to achieve compliance. The primary legal considerations for building performance standards relate to state law preemption, enforcement and penalties, the regulation of small buildings, and displacement and equity.[84]

State Law Preemption

A number of state laws may preempt local building performance standards. *Statewide building codes* are of particular concern, as some include requirements for *existing* buildings that could preempt, either expressly or through field preemption, a local building performance standard.

State air pollution control laws may also preempt some building performance standards when the standards are structured as limits on GHG emissions.[85] Some states, like Michigan, allow local governments to enact emissions requirements that are more stringent than state requirements.[86] Others, like California, have far-reaching state air pollution control laws and delegate a significant amount of authority over air pollution control to air districts covering several municipalities.[87] Still others, like Virginia[88] and Minnesota,[89] preempt local air pollution control requirements altogether.

Though the issue is less well explored, *state public utility laws and regulations* may also have preemptive effect. As will be discussed in chapter 5, states retain authority over most aspects of energy distribution law and regulation. In theory, a clear line separates the oversight of energy generation and distribution from building energy use. In practice, however, state energy requirements can be far-reaching, and cities must ensure that their requirements don't conflict with or tread into that "field." States may also have laws and rules relating to *building occupancy*, *safety*, *tenant protections*, and other issues. These could include requirements relating to energy efficiency, insulation, fire prevention, ventilation, and other areas affecting building decarbonization.

Fees, Taxes, and Penalties

Where a local government wishes to impose a monetary fee or penalty for noncompliant buildings, it may seek to do so through fees, taxes, and penalties. As discussed in chapter 1, many local governments have the authority to impose *fees*,[90] whereas fewer have the authority to impose *taxes*.[91] The legal risk here is that a local government might impose what is intended to be a fee on carbon or energy use in buildings only to have a court deem the charge a tax for which the local government does not have the authority to implement. As a third option, a local government might consider a *penalty* on excess building emissions or energy use. States frequently delegate authority to municipalities to assess penalties.[92]

Small Buildings

At present, most building performance standards apply only to large buildings. New legal questions will emerge as cities look to control the emissions or energy use of smaller buildings. As discussed earlier, collecting information on emissions or energy use from buildings can give rise to privacy concerns. Requirements for small or single-family buildings may also cause practical issues—for example, how will single-family homeowners afford upgrades or navigate available options?—that will require careful consideration to avoid legal roadblocks. In some states, for example, municipal tax dollars may not be used for improvements on private property, meaning a local government may not be able to finance or provide building retrofits itself.[93] In many others, the distinction between taxes and fees may dictate available options.[94] Some municipalities need state authorization to impose a new tax to fund building improvements, whereas others may be unable to do so at all.[95]

Gentrification, Displacement, and Rising Rents

Policies aimed at upgrading building stock may have equity implications such as the potential for increased rents, neighborhood gentrification, and the displacement of residents. These issues came to a head in the negotiations leading to the enactment of New York City's building performance standard, Local Law 97. At issue was New York's state-level rent stabilization regime, which allowed owners of rent-stabilized residential units to increase rents by up to 6 percent per year to recover the costs of "major capital improvements."[96] If applicable to buildings with rent-stabilized units, Local Law 97 could have led building owners to comply by completing major capital improvements and passing the costs on to low- and middle-income tenants. The final law addressed this risk by providing buildings with one or more rent-regulated units with a list of prescriptive measures that, if undertaken, would establish compliance.[97]

PRESCRIPTIVE BUILDING REQUIREMENTS

The category of "prescriptive" building requirements captures a wide range of requirements that specify building upgrades. In some instances, prescriptive building requirements or options act as alternatives to building performance standards for some or all buildings. This is the case under Reno's building performance standard, which offers several prescriptive compliance pathways, including LEED certification and participation in a utility building retuning program,[98] and New York City's Local Law 97, which allows residential buildings in which at least 35 percent of the building's units are affordable apartments to comply by completing a list of prescriptive measures like insulating pipes and tanks, repairing certain appliance components, and adjusting temperature set points for heat and hot water.[99]

Other prescriptive requirements can stand on their own. A *retrocommissioning requirement* requires an assessment and retuning of building systems, particularly in large buildings, to ensure that existing building equipment works in the most energy-efficient way possible. Depending on the building system, retrocommissioning may include minor improvements to equipment (say, replacing a valve) or adjustments to individual pieces of equipment so that they work more optimally as part of a system but stops far short of capital improvements or whole-equipment replacements.[100] The terms *green roof* and *cool roof* describe a range of roof surfaces and uses that can reduce a building's energy use (by providing extra insulation or repelling the sun's rays on hot days), mitigate the impacts of urban heat islands, or manage stormwater. Policies to encourage these beneficial uses of rooftop space can use a number of mechanisms. New York City, for example, has several overlapping policies to increase these roof spaces: a building code provision requiring new roofs to be covered in green roofing or be fitted for renewable energy generation,[101] a state-level tax credit for green roofing, and a local initiative offering a cool roof coating to select buildings.[102]

Prescriptive building requirements, like other forms of building energy or GHG requirements, may conflict with and be preempted by state or federal law. Certain forms of prescriptive requirement may not be authorized under applicable state law, and accountability measures will need to be tailored to the frameworks of applicable state laws. With those caveats in mind, many building requirements can be lawful in various local jurisdictions.

TRIGGERS FOR THE DECARBONIZATION OBLIGATIONS OF EXISTING BUILDING

Boulder's rental efficiency standards illustrate how a city might establish regulatory triggers to capture smaller buildings. Boulder's standards apply to residential rental buildings and must be achieved before a building owner obtains a rental license.[103] Austin's Energy Conservation Audit and Disclosure Ordinance has triggers for both sales and rentals.[104] San Francisco's Residential Energy Conservation and Residential Water Conservation Ordinances[105] apply before a building's sale, requiring both an inspection and potential mandatory upgrades. San Francisco's ordinances apply only one time, have relatively lenient standards, and are now quite outdated.[106] However, this model of requiring inspections, disclosure, or building upgrades before a building's sale may be useful. A city might also consider periodic requirements, such as inspections or upgrades every five or ten years—frequently enough to drive emission reductions but not so frequent as to overburden the owners of smaller buildings.

PROPERTY-ASSESSED CLEAN ENERGY FINANCING

Property-assessed clean energy (PACE) financing offers low-interest loans to finance energy conservation and renewable energy retrofits to existing private buildings. PACE loans tie to the property itself, rather than the borrower, and loan amounts are paid back as a line item on property tax bills.[107] This financing structure allows the loan to continue in effect after the sale of the property, meaning that subsequent owners share in the cost of building improvements. Longer repayment periods make it more likely that utility bill savings will free up cash to pay down the loan. PACE financing generally must be authorized by state law, pursuant to which local governments can enact the financing mechanism through the creation of "districts" through which property owners may opt in to the program. PACE is offered at the state or local level in at least thirty-three states. As applied to single-family homes, PACE has been argued to undermine equity and housing stability, as it can saddle low- or middle-income homeowners with significant debt secured with a lien on their homes. Los Angeles County, facing lawsuits alleging the county had not properly overseen predatory PACE lenders, discontinued its residential PACE program in 2020. Some cities may choose to limit PACE financing to commercial properties (often called "commercial PACE," or "C-PACE").

CLOSING MARKET AND KNOWLEDGE GAPS WITH LEGAL TOOLS

Building emissions policy can be complicated by market failures that leave no party with the economic incentive to retrofit a building or build sustainably from the start. One such gap is the so-called split-incentive problem, a mismatch in the incentives for building owners and tenants of existing buildings: a landlord is generally more able to retrofit a building to conserve energy, but the cost savings will largely be passed along to a tenant who pays the utility bills.[108] (A gross lease, under which the landlord pays for utilities, also does not fully address the problem, as tenants are not incentivized to conserve energy.[109]) Legal tools are developing to help address the split-incentive problem, including the use of green leases, which better allocate incentives between landlord and tenant both to invest in retrofits and share in utility bill savings over time.[110] Also termed "energy-aligned" leases, these agreements are entered into by private parties—not the city—but the city can play an active role in developing templated energy-aligned language, educating landlords and tenants about the availability of such language, and soliciting stakeholder input on how to facilitate the relationship between landlords and tenants.[111]

Similarly, in new buildings, energy cost savings are realized on monthly utility bills over the life of the building, leaving a developer with little economic incentive to invest in a sustainable building at the time of construction. However, the city can again play a role in encouraging builders to build low-emission or energy-efficient buildings, in this case through affirmative requirements such as building codes and information-sharing among stakeholders.

BUILDING WASTE AND EMBODIED CARBON

While cities have typically focused on GHG emissions from building operations, policy-makers are increasingly considering the emissions associated with building construction. Many building materials are carbon intensive, particularly steel, iron, and concrete,[112] and some features meant to reduce operational emissions may involve the use of materials that drastically increase emissions during construction.[113] Frameworks for regulating the embodied carbon in building materials are in the very early stages of development. Cities without delegated building code authority are unable to pass local construction requirements with respect to low-carbon building materials. In many states with statewide building codes, even local laws that are not couched as building code requirements can be preempted by the statewide code.[114] At the same time, many cities regulate construction waste in some way, and these requirements can be fine-tuned to require that discarded building materials do not go to landfills or incentivize the use of materials with less embodied carbon. Pasadena, California, for example, "requires that certain construction and demolition projects divert at least 75 percent of waste either through recycling, salvage or deconstruction."[115] Construction waste regulations, if properly drafted, would have the benefit of not being building requirements that could be subject to state preemption.

MUNICIPALLY OWNED AND OPERATED BUILDINGS

Cities can act as market participants to reduce GHG emissions from buildings. Acting to electrify, retrofit, power with renewables, or otherwise address emissions from city buildings avoids many of the thorny legal questions that can arise when seeking to take on private buildings, while also proving the market for green building products and services. Retrofitting city buildings with electric heat pumps, for example, may show that the technology is cost-effective and feasible, easing the political path to later policies aimed at private buildings. Similarly, "if local construction and design firms have LEED-certified personnel for public projects, they will bring green building expertise and awareness to their private sector projects as well."[116] Many cities introduce climate policies by implementing them at the government level first. Addressing GHG emissions and energy use in new and existing buildings provides another opportunity for city climate leadership.

EQUITY AND REDUCING BUILDING EMISSIONS

Building policy for GHG emission reductions is deeply intertwined with questions of equity. Building policies have the potential to exacerbate historical and ongoing practices of racial exclusion like redlining. Buildings constructed or retrofitted to be energy efficient or use low-emitting electrical systems can improve air quality and thermal comfort in affordable housing stock but also risk leading to "green gentrification." Stakeholder engagement is critical to understanding the potential impacts of policies aimed at reducing building emissions on frontline communities and how to tailor policies to benefit low-income and racialized residents.[117]

4

REDUCING TRANSPORTATION-RELATED GREENHOUSE GAS EMISSIONS

With transportation accounting for 29 percent of all GHG emissions in the United States[1] and for the largest share of GHG emissions in many cities,[2] reducing transportation emissions is a critical component of any city's climate action plan. Moreover, measures that reduce GHG emissions from vehicles also reduce tailpipe pollutants like nitrogen oxides, sulfur dioxide, and particulate matter, which have many negative health impacts, including increased incidences of asthma and premature death among those who live in heavily polluted areas. Some transportation policies also bring added benefits like safer and more pleasant streets for pedestrians, revitalized downtown areas, and fewer traffic fatalities.

This chapter looks at three interrelated approaches to reducing a city's transportation-related GHG emissions:

1. Scaling up the adoption of electric vehicles (EVs): Many policy tools for increasing EV use are more appropriately deployed at the federal or state level, but cities still have legal tools available to them to encourage EV adoption.
2. Limiting traffic in the center of a city: Congestion pricing and low-emission zones (policies we refer to jointly as "low-traffic

zones") come in many variations and can involve bans and charges for various types of vehicles.

3. Catalyzing mode shift: Investments in public transit, bicycle and pedestrian infrastructure, and adding micromobility or other last-mile transportation options are means of encouraging city residents and service providers to shift their modes of travel.

Limiting traffic through LTZs and catalyzing mode shift are both means of reducing vehicle miles traveled. For many cities, reducing vehicle miles traveled and increasing EV adoption are complementary; that is, reducing transportation GHG

A NOTE ON EQUITY

Policies aimed at reducing transportation GHG emissions have the potential to remake neighborhoods and to either address or perpetuate inequities like exposure to elevated levels of air pollution and inadequate access to transit. Choices about where to deploy EV fleets and site charging stations, offer incentives for EV ownership, expand transit access, and build cycling and pedestrian infrastructure can help create a more equitable city that limits the burden of pollution on low-income communities and communities of color and enhances economic opportunity by connecting all city residents to transit. They can also do the opposite.

No city leader or group of city leaders can be fully aware of how a strategy or plan to decarbonize a transportation system will affect neighborhoods, communities, and individual lived experiences. For this reason, some state and local environmental and land use planning processes require public input through a hearing or public comment period. Even where the applicable laws do not require such input, city laws and programs aimed at reducing GHG emissions are increasingly building in opportunities for public participation through appointed working groups and advisory boards, a strategy that can be deployed in incorporating equity considerations into transportation planning.

emissions to the level needed to achieve significant decarbonization requires eliminating as many vehicle trips as possible in favor of more sustainable modes of travel like walking, cycling, and taking public transit and ensuring that necessary vehicle trips are made in EVs.

FEDERAL TRANSPORTATION LAW

Significant parts of U.S. transportation policy are set at the federal level, including national standards that govern fuel economy and emissions from motor vehicle engines. Automobile manufacturers, among others, argue there are good reasons for this centralized policy-making, including avoiding differing vehicle standards across states. It does, however, limit the ability of cities to set policy relating to vehicles, particularly engine and emissions requirements.

Federal law can preempt local laws aimed at reducing transportation GHG emissions. The Energy Policy and Conservation Act (EPCA) and the Clean Air Act (CAA), in particular, restrict local governments from setting requirements that vehicles driven in a city meet specified fuel economy or emissions standards. EPCA expressly preempts any state or local "law or regulation related to fuel economy standards or average fuel economy standards for automobiles,"[3] and more than one federal court has held that state and local laws requiring that classes of vehicles like taxis or trucks be hybrids or use other clean engine technology are preempted by EPCA.[4] The CAA preempts "any [state or local] standard relating to the control of emissions from new motor vehicles or new motor vehicle engines,"[5] and federal case law has similarly extended this language to preempt a local requirement that private fleet operators purchase low-emission vehicles.[6] While this case law has not caught up to EV technology (it currently assesses hybrid and compressed natural gas engines), the clear implication is that fuel economy or air pollution standards requiring vehicles to be electric would also be preempted under these lines of case law.

States and, if authorized, local governments have the authority to offer incentives aimed at increasing EV uptake or reducing the use of vehicles with traditional internal combustion engines, and incentives to purchase high-efficiency, low-emission, or zero-carbon vehicles will not be preempted by EPCA or the CAA. However, the line between local incentives and mandates isn't always clear. Incentives cannot be "so coercive as to indirectly mandate"[7] the purchase or use of EVs or other lower-emission or more fuel-efficient vehicles. Courts have identified some municipal incentives that stay on the nonpreempted side of the fine line. For example, a Dallas program that allowed taxis powered by compressed natural gas to cut to the head of the pickup line at the airport was upheld.[8] A program in King County, Washington, that set aside fifty taxi licenses for hybrid vehicles also was upheld as a permissible incentive.[9] But where a local incentive presents "an offer which cannot, in practical effect, be refused,"[10] it may be considered a de facto mandate and preempted.[11]

FEDERAL CONSTITUTIONAL LAW

The U.S. Constitution grants the federal government authority to regulate interstate commerce.[12] This provision, the Commerce Clause, is understood to have a silent, or "dormant," aspect that guards against protectionism and prohibits states and local governments from regulating in a way that discriminates against interstate commerce.[13] Vehicles are bought and sold in all fifty states, and heavy-duty vehicles often carry goods for sale across state lines; when local governments regulate with respect to transportation, they must take care to avoid violating the Dormant Commerce Clause (DCC).

There are two relevant DCC tests for local transportation policies. First, a law that significantly discriminates against interstate commerce will, with a few exceptions, be considered per se invalid.[14] For example, strict limitations on out-of-state vehicles without such limitations on in-state vehicles would likely violate the DCC. Second, state and local laws that have only "incidental effects on interstate commerce" will be upheld where the "statute regulates evenhandedly . . . [and] unless the burden imposed on such commerce is clearly excessive in relation to the putative local benefits."[15] (This is referred to as the "Pike balancing test.") In other words, a court will weigh the purported benefits of a transportation policy—perhaps reduced traffic, improved local air quality, or increased safety—against the burden that some vehicles may face limitations or costs on their routes.

MARKET PARTICIPANT EXCEPTION

While cities are bound by limitations of federal statutory and constitutional law in acting as governing entities and regulators, they have far more latitude to act as direct market participants—that is, to spend their own money and use their own property. The "market participant exception" has been applied in cases related to the DCC, EPCA,[16] the CAA,[17] and the Federal Aviation Administration Authorization Act[18] to allow or consider allowing municipalities to make policy determinations that leverage their own purchasing power. As one federal circuit court put it, "Actions taken by a state or its subdivision as a market participant are generally protected from federal preemption."[19]

ENVIRONMENTAL REVIEW

Several states, such as California and New York, have environmental review statutes requiring state actors, including municipalities and municipal agencies, to assess the environmental impacts of their actions.[20] These environmental review requirements offer an easy legal "hook" that project opponents can use to challenge an effort to reduce transportation emissions in court.[21] While the risk of litigation cannot be ruled out (in some places it may be almost a certainty), a city's careful adherence to substantive and procedural requirements of these environmental review laws can help ensure an effort to build EV chargers, transform a city streetscape, or invest in public transit survives legal challenges.

LEGAL CONSIDERATIONS FOR THE EXPANSION OF ELECTRIC VEHICLES

Because a city would almost certainly be preempted from mandating that new automobiles be electric, and because fiscal limitations would likely prevent cities from offering direct incentives for EV purchases, municipalities can encourage EV adoption by building out robust charging networks throughout the city.

Local efforts to scale up EVs can come under EPCA and CAA preemption scrutiny. As discussed, incentives cannot be mandates in disguise. But a number of programs have survived judicial review: the program that allowed taxis powered by compressed natural gas to cut to the head of the airport pickup line in Dallas;[22] the fifty taxi licenses set aside for hybrid vehicles in King County;[23] and (at the state level) a range of financial incentives for EV purchases. Cities also generally have some authority over public parking and curb space and so may be able to offer incentives such as priority parking at EV charging stations or free or subsidized charging. Salt Lake City, Utah, for example, offered free charging for a time.[24] While EV charging is a rapidly developing area, adjusting parking prices and requirements for street, lot, and garage parking to influence traffic patterns in a city is not new.

Some incentives to install EV chargers might be aimed at employers and real estate developers. Cities have also offered rebates for purchases of low-emitting vehicles to public agencies and incentives for purchases of low-emission vehicles by low-income drivers. This latter approach can help alleviate inequities between those who can afford to purchase new EVs, and thus reap the benefits to local air quality of phasing out vehicles with internal combustion engines, and those who cannot.

ELECTRIC VEHICLES AND STATE PUBLIC UTILITY LAW

EV chargers do something that has long been the sole right and responsibility of electric utility companies: sell power. The long-standing stature of electric utilities as regulated monopolies has given rise to questions about how the sale of electricity directly to drivers should work. In most states, state law exempts EV chargers from the definition of *public utility*,[25] allowing third parties to install public charging stations without being subject to utility regulation. For example, Kentucky's Public Service Commission found that EV charging stations need not be regulated by the Commission because charging stations serve only a "specific, defined class of persons" (i.e., those who own EVs) rather than the broader public.[26] A handful of other states have yet to determine whether EV charging stations or providers should be regulated as utilities.

Another significant variable is whether public utilities can earn a return on their investments in EV charging infrastructure.[27] California, for example, allows utilities to pass charging installation costs along to ratepayers, and all three electric utilities in the state have responded by building chargers.[28] Missouri, Michigan, and Kansas have taken the opposite tack and do not allow utilities to recover costs for installing EV chargers, in part because of a belief that non-EV owners should not subsidize chargers for EV owners.[29] While cities do not make these determinations, the decisions underscore the significant role that public utility companies play in building out a city's network of EV chargers.

In any regulatory scenario, electric utilities are major stakeholders and may be tasked with building out the charging network, or they may need to weigh in on their ability to meet an increased demand for electricity. These matters are complex and require negotiated agreements among the city, state regulators, the utility, and other stakeholders that comport with federal and state laws and regulations. Cities with publicly owned electric utilities may be better able to smooth the relationship between the city and the utility. For example, Seattle's publicly owned utility runs the Drive Clean Seattle program, which allows residents to repay the cost of installing an EV charger at home through their electric bills.[30] In Texas, Austin Energy (also publicly owned) has several EV programs, including a rebate for charger installation, flat-rate charging for its large network of chargers, and e-bike incentives.[31]

ELECTRIC VEHICLE CHARGING ON PRIVATE PROPERTY

In addition to installing chargers in public rights-of-way, many cities require that new buildings or new parking locations include EV chargers or be wired for eventual installation. Several state and local building codes include EV charging or EV-readiness requirements.[32] Imposing these requirements raises a set of legal questions.

For one thing, many cities lack building code authority: at least twenty states set building code requirements at the state level with limited opportunity for many or any municipal amendments.[33] Where states retain all or most building code authority, cities are not only precluded from passing local building code provisions for EV charging but also need to take care not to enact a requirement that would be preempted by the state building energy code. In other words, such a city could not simply pass a construction requirement outside the building code to avoid preemption.[34] (For more on building codes, see chapter 3.) These cities might seek to engage with state lawmakers to update the state-level code to include EV charging equipment or to pass a stretch code (an alternative building energy code set at the state level that municipalities may choose to adopt).[35] As an alternative, a municipality might offer incentives to private building or property owners to install EV charging infrastructure. Incentives could be financial, in the form of a zoning bonus, or something else.

Cities can also use their zoning powers as a source of authority to require or provide opportunities for expanded EV charging on private property. First, they can include *direct requirements* that parking spaces for a new building or development include EV charging. Cities can add to the minimum or maximum requirements for parking a condition that a certain number of spots include charging stations or be wired to be EV-ready to accommodate future chargers. Salt Lake City, Utah's zoning code, for example, requires one EV charging space for every twenty-five spaces in multifamily buildings.[36]

Second, a city can revise its zoning code to *clarify requirements* for EV charging stations. Developers may be stymied in building new EV chargers by unclear or overly burdensome zoning requirements. The City of Chelan, Washington, for example, updated its code to remove barriers to installing small chargers, clarifying that they are permitted in all zoning districts, and established a process for the approval of large charging stations in commercial and industrial zones.[37] Petaluma, California, used its zoning authority to disallow new gas station development and to encourage the build-out of EV charging infrastructure.[38]

Finally, a zoning code can offer *density bonuses or other incentives* to a developer in exchange for building EV parking and charging in conjunction with a new project. Where a direct requirement is not lawful (as where it would conflict with or be preempted by a state building code), zoning authority can serve as a powerful tool.

SITING IN THE PUBLIC RIGHT-OF-WAY

In siting chargers on sidewalks or public roadways, cities need to (a) contract with the party (utility or otherwise) operating the charger and selling electricity to vehicle owners and (b) adhere to any street design or safety requirements. Cities looking to augment their charging networks with direct-current (DC) rapid chargers, which charge a vehicle in a relatively short amount of time (at least 60 miles in twenty minutes of charging), need larger locations in which to site them (a DC rapid-charging station is more similar in size to a gas station than to a curbside charger). As cities survey available municipal land for charging infrastructure, they must consider permissible uses of such land, and charging stations must comply with local land use designations, including public trust protections.[39] Any taking of private property to site EV charging infrastructure requires just compensation to the property owner.[40]

Of course, drivers don't stay within municipal bounds, and EV drivers may be stymied by a lack of charging options in other municipalities (for example, a commuter may drive to work in a neighboring city that does not have widespread charging). For this reason, cities could consider working across a metropolitan region to build a far-reaching system of chargers. Because cities derive their authority from state delegations, a state statute must authorize intergovernmental cooperation among localities. More than forty states authorize some form of cooperation among local governments;[41] some of those states require state approval for interlocal agreements.[42] In any event, cities pursuing this path must look carefully at the limitations of any such authorization. The federal government also plays a role here, including by offering funding for projects, such as through 2021's Bipartisan Infrastructure Law.[43]

Cities have additional legal tools at their disposal for facilitating expanded EV travel. They might allow certain public areas—curb space or parking spots—to be set aside for charging EVs. Cities and states might also consider establishing a new traffic offense that penalizes vehicles (EVs and non-EVs) parked in EV charging spots without using the charger. In many states, the authority to create or enforce a traffic offense is retained at the state level, but state governments may be willing to work with municipalities to enact such a state law, as Colorado did.[44] A local ordinance could also generally be used to develop a parking offense or to facilitate signage identifying EV charging zones.

REGULATING TRANSPORTATION NETWORK COMPANIES

Who regulates transportation network companies (TNCs) like Uber and Lyft? The answer to this question depends on state delegations of authority to municipalities or lack thereof. Many states delegate at least some oversight of taxicab companies to cities, though these delegations sometimes come with limitations (e.g., in the state of Washington, local laws relating to taxicabs must relate to a short list of parameters such as taxi business regulation and "safe and reliable taxicab service,"[45] restrictions that may not allow for some emission-limiting requirements). TNC regulation is also patchwork, with some states authorizing municipal-level laws and others retaining authority at the state level, possibly under the control of the state public utility commission. In some places, the legal regimes for taxis and TNC vehicles are different, further complicating their regulation.

CONSUMER MEASURES

A few cities have taken measures to curb the sale of gasoline in their communities. While not a direct tool for reducing in-city transportation GHG emissions, these cities are betting on gasoline's demise. The city of Cambridge, Massachusetts, enacted a warning-label requirement for gas pumps; the warning must "explain[] that burning gasoline, diesel and ethanol has major consequences on human health and on the environment, including contributing to climate change."[46] Petaluma, California, took the step of zoning out new gas stations (and, just as critically, enacting new zoning measures to encourage the construction of low-GHG fuel infrastructure like EV charging and hydrogen fuel cell stations).[47] Both cities' measures should be taken with a grain of salt, however—neither actually makes it harder to purchase gasoline. Cambridge's law is only a labeling requirement, and Petaluma's can be viewed as a land use planning decision. It is based on the city's assessment that all residents—including residents of homes yet to be built—are able to reach at least one gas station within five minutes, obviating the need for more.[48]

MUNICIPAL FLEETS AND CHARGING INFRASTRUCTURE

From the perspective of federal law, there is little stopping cities from investing in EVs for their own fleets. The market participant exception addresses preemption and Dormant Commerce Clause concerns, and the legal considerations around siting are generally less relevant in this context. Many cities have low-emission programs or goals for their municipal fleets. Similarly, some cities choose to purchase or contract for their own chargers. Cities are bound by all the usual state and local procurement, contracting, and other related rules.

ELECTRIFYING HEAVY-DUTY VEHICLES AND FREIGHT

Electrifying heavy-duty vehicles and freight brings with it a different set of policies and legal considerations from that of electrifying light-duty cars and trucks. Much of the law and policy for heavy-duty vehicles comes, again, from federal statutes and, to a lesser extent, state legislation. Still, cities can play a role in reducing GHG emissions from heavy-duty vehicles and freight by purchasing and piloting EV programs for buses and municipal trucks like garbage and recycling trucks.[49] Cities can also install EV chargers for large freight vehicles and electrify any port or other freight infrastructure within its control.[50] And cities can help mitigate the harmful impacts of truck traffic, like local air pollution, by soliciting and considering public feedback on routes and depot siting for these large vehicles.

A NOTE ON AUTONOMOUS VEHICLES

While EVs and autonomous vehicles (AVs) are not synonymous, some AV proponents advocate that any future AVs in widespread use should be electric and that the widescale deployment of AVs could be part of a comprehensive plan to reduce GHG emissions by acting as a last-mile connection to public transit, facilitating package deliveries, and reducing the need for households to own private vehicles.[51] Introducing AVs to a city's transportation system will require a robust new regulatory regime. While vehicle standards will likely need to be promulgated at the federal or state level, local governments will likely play a role in enacting laws aimed at safety, the rules of the road, liability, and other matters. It is not yet clear whether AV technology can help cities decarbonize the transportation sector.

LOW-TRAFFIC ZONES

Congestion pricing and low-emission zones (LEZs) offer two primary approaches to limiting vehicle miles traveled in a city center. As explained by the Carbon Neutral Cities Alliance, "car-free and low-emission zones are 'travel demand management' approaches that use different tools—e.g., a ban and a price—to change driving behaviors" in a defined geographic zone.[52] Congestion prices, LEZs, and full-out bans on vehicles give rise to similar legal questions. We define these approaches together as *low-traffic zones* (LTZs): bounded geographic areas in which reductions in vehicular traffic are achieved or attempted through legal and policy approaches.

FEDERAL PREEMPTION OF LOCAL LOW-TRAFFIC ZONES

As with many city-led transportation policies affecting vehicles, EPCA and the CAA have the potential to preempt LTZs. The key question is generally whether the fee, ban, or other mechanism meant to limit traffic in the zone is a de facto mandate, preempted by EPCA and the CAA, or an incentive, and thus not preempted.

What does the distinction between de facto mandate and incentive mean for LTZ policy? It's not entirely clear yet. LTZ strategies like LEZs and congestion pricing should be able to employ incentives differentiating among vehicles based on engine technology, but the law draws no clear lines. A city could credibly, and perhaps successfully, argue that congestion charges or LEZ entry fees that vary based on type of engine are an incentive. Moreover, a city needn't limit itself to monetary incentives or disincentives—drivers might be moved to adopt EVs in greater numbers by nonfinancial inducements over which a city may have clearer authority: access to parking, charging stations, or priority loading and unloading zones within the LTZ, among other things.

The Federal Aviation Administration Authorization Act also has the potential to preempt city or state LTZ policies, as it preempts state and local laws relating to the "price, route or service of any motor carrier . . . with respect to the transportation of property."[53] Still, many cities enjoy fairly wide latitude (depending on state law) to set truck routes, tolls, or other traffic restrictions based on vehicle weight (this latter item is expressly carved out from potential preemption by the act and may provide a tool for limiting heavy-polluting vehicles like large diesel trucks but not for specifically prioritizing EVs). The Federal Aviation Administration Authorization Act also has a market participant exception.

CONSTITUTIONAL CONCERNS RELATING TO LOW-TRAFFIC ZONES

LTZ policies can also face significant questions about potential DCC limitations. LTZ laws and policies can have at least an "incidental" effect on interstate commerce—they can affect the transportation of goods across state lines—but they generally can be structured to avoid facially discriminating against interstate commerce and to satisfy the Pike balancing test mentioned earlier.[54] Local benefits of LTZs include reductions in traffic and emissions and the protection of health and safety, all of which could be referenced to justify an LTZ that does not overtly favor vehicles from one state over another.

AUTHORITY TO TOLL

Congestion pricing adds a particular wrinkle for municipalities: where does a city get the authority to impose a toll? This question must be answered at both the federal and state levels.

Any road considered a federal-aid highway—that is, a road eligible for federal funding—must comply with U.S.C. Title 23 ("Highways"), and approval is generally required from the Federal Highway Administration to enact a congestion charge. The Administration can facilitate state and local projects that study, pilot, or implement congestion pricing strategies, but it also can hold up their approval. In particular, cities must comply with the National Environmental Policy Act.[55]

State law controls congestion pricing and other tolls on roads not considered federal-aid highways. Some states, like Oregon,[56] generally allow municipalities to collect tolls on the roads each municipality manages. Others, like New York,[57] reserve toll-setting authority for the state.[58] The state of Washington allows localities to create "transportation benefit districts" that can charge tolls approved by "a majority of the votes in the district . . . at a general or special election."[59] State laws can also limit how tolling revenues are spent. The conventional wisdom is that congestion pricing, if done equitably and to reduce GHG emissions, should be paired with investments to improve transit, bicycle, and pedestrian infrastructure. If a city's use of tolling revenues is restricted, it must spend funds consistent with those limitations.

CLOSING A ROAD

Cities often have clearer authority to close a road to vehicular traffic entirely than to enact a cordon pricing regime or impose conditions on which types of vehicles can enter an area, making road closures an important LTZ strategy. Municipalities often have broad, state-delegated authority to regulate street traffic, granted either expressly through state statute or implicitly as part of its police power. Courts in Idaho,[60] Connecticut,[61] and elsewhere have held that city closures of streets to all but cyclists and pedestrians fall within applicable municipal authority, including to "advance economic, aesthetic and safety-related goals."[62] Street closures grew in popularity in 2020 during the onset of the global COVID-19 pandemic and related shutdowns; many cities responded to the pandemic by closing streets to vehicles and "opening" them for exercise, recreation, and essential workers' commutes.

PRIVACY CONSIDERATIONS

Privacy and data security can prove thorny for LTZs and congestion pricing programs in three broad ways. First, cameras called *automated license plate readers* (ALPRs) are often used to monitor tolled roads and congestion zone boundaries to collect tolls electronically. Several states have enacted laws governing ALPRs: who can access the data and why and how long the data can be stored.[63] State statutory and case law can play a role here, with some courts holding that a person's location data compiled from cumulative ALPR readings implicates protected privacy interests.[64] Some state laws, such as the California Consumer Privacy Act,[65] also set more generally applicable data protection requirements that must be considered.

Second, *onboard payment mechanisms* must have some way to track when a vehicle crosses the cordon or toll point. Some *cordon pricing* and *area-wide pricing* schemes make use of more immediate location data, such as through global positioning systems (GPSs), to charge drivers within a set zone. Collecting real-time GPS data gives rise to significant questions about privacy, some of them not yet entirely settled. Sometimes, state privacy laws require clarification to ensure that GPS data collected for tolling purposes is covered by these laws such that drivers' privacy is protected.[66]

And third, some cities require *trip data from for-hire vehicle companies* to assess the activity of for-hire vehicles in a cordon zone (relevant where cities have implemented or are considering implementing *fleet pricing* for for-hire vehicles within a cordon zone). These data are subject to federal, state, and local data security requirements, and data collection policies have in some instances led to litigation.[67]

DEFINITIONS

Cordon pricing programs charge vehicles a toll (often once a day) upon entering a set geographic zone.

Area-wide pricing regimes charge per mile traveled within a cordon pricing zone.

Fleet pricing is a form of congestion pricing applied to one or more categories of vehicles. New York City[a] and Chicago[b] have fleet pricing in the form of a rider surcharge for for-hire vehicle rides (taxicab rides and rides provided through app-based companies like Uber and Lyft).

[a]N.Y. Comp. Codes R. & Regs. tit. 20, pt. 700 (2019).

[b]Chicago, Ill., Ord. O2019-8527 (Nov. 26, 2019).

CATALYZING MODE SHIFT

It's not enough for a city to limit vehicle use and replace those vehicles still in use with EVs; it must provide alternative modes of travel and encourage their adoption. The barriers to doing this are less legal—though certainly authority over road design and public transit services vis-à-vis the state must be sorted out—than financial and political. Who will pay for new bike lanes or bus routes, and how? What entrenched interests may continue to push car use? From a legal perspective, one of the simplest ways to increase public transit use is to make it free to riders, as Kansas City, Missouri, did in 2019.[68] Of course, many cities cannot forgo fare revenue.

Several legal tools within municipalities' land use authority can make walking, cycling, and taking transit more appealing. Transit-oriented development generally involves planning higher-density development that offers a range of residential and commercial uses within walking distance of a transit hub.[69] Aligning city plans and zoning codes to support this form of development can help eliminate vehicle trips, and associated GHG emissions, by people who live near shopping, restaurants, and work or their train to the office. Transit-oriented development zoning provisions often have incentives like reductions in the required parking minimums (as in Chicago)[70] and extra floor area allowances to provide pedestrian-oriented amenities like parks and retail.[71]

Revisions to local parking requirements can also help reduce driving. Many zoning codes contain *parking minimums*: a required number of parking spaces, which varies by development use and density, that must accompany new buildings. Often, this parking is unneeded or encourages more driving.[72] Cities can use zoning codes to reduce parking minimums.

Doing so creates two benefits that can yield emission reductions: it can encourage some travelers to use modes of travel other than cars because they know it will be difficult to find a parking space, and it can create more walkable districts where buildings are not separated by enormous parking lots. Some cities have enacted *parking maximums*, which "establish an upper bound for the number of spaces allowed for a specific use, thus controlling the amounts of land and impervious surface associated with parking."[73] Several cities have set parking maximums in their zoning codes.[74] A third alternative is to allow or require *parking in-lieu fees*, for which a developer pays into a fund "used to pay solely for one or more large parking developments that serve an entire district."[75]

Other legal and policy tools can be used to incentivize biking and walking, particularly when used as part of a comprehensive plan to build a network of nondriving modes of transportation. With *alternative pedestrian routes*, a city uses its zoning powers to require developers to build pedestrian pathways between new buildings and existing sidewalks or pedestrian paths.[76] *Complete Streets* refers to a broader range of tools underlying a policy that prioritizes "the safe and adequate accommodation . . . of all users of the transportation system, including pedestrians, bicyclists, public transit users, children, older individuals, individuals with disabilities, motorists, and freight vehicles."[77] In many cities, a central component of Complete Streets is a Vision-Zero goal—that is, a goal to eliminate fatalities caused by vehicles.[78] Safer pedestrian and cycling options mean that more people are likely to walk and bike, linking the goals of increased street safety and reduced transportation GHG emissions. A complete-streets policy necessarily involves a range of road safety strategies relating to a city's authority over land use, streets, and traffic.

MICROMOBILITY

Companies offering bikes, e-bikes, and e-scooters have spread across the country. It is not yet clear how effective these modes are in reducing overall GHG emissions, at least over a bike or scooter's life cycle,[79] but there is at least some reason to think that they may lead to some reductions. The legal landscape for micromobility was murky at the start, particularly for "dockless" micromobility, a business model in which shared bikes and scooters don't have designated storage locations and can be left almost anywhere in the public right-of-way. Dockless micromobility companies initially dumped hundreds of bikes and scooters on unsuspecting cities, leaving the cities to figure out a regulatory scheme that covered various legal issues relating to public rights-of-way; insurance, indemnity, and bonding; safety; parking; data sharing; customer service; device types; and permitting.[80] While best practices have emerged, so too has litigation relating to the Americans with Disabilities Act[81] (alleging discrimination against people with disabilities who might trip over bikes and scooters left on the sidewalk)[82] and the sharing of customer data with the host city.[83]

VOLUNTARY STRATEGIES FOR LAST-MILE DELIVERY

Perhaps it makes sense to end this chapter by addressing last-mile delivery. Last-mile delivery is a ripe area for mode shift, particularly for a city's denser areas. Not only do delivery trucks emit GHGs and other pollutants but also many city streets simply cannot handle them. While delivery trucks are good candidates for electrification, small cargo e-bikes are another green option.

A few U.S. cities, including New York[84] and Seattle,[85] have allowed logistics companies to pilot cargo e-bikes for package deliveries.

The city of Santa Monica, California, developed an even more comprehensive last-mile delivery program. Designating a one-square-mile "zero-emission delivery zone," the city worked with businesses, technology start-ups, and residents to pilot micromobility last-mile delivery options, medium- and heavy-duty electric delivery trucks, and curb zones offering e-truck charging.[86] What's notable is that in developing this first-in-the-nation zero-emission zone, Santa Monica used essentially none of the legal tools described throughout this chapter. Its strategy implicitly acknowledges the city's limited legal authority to require deliveries by zero-emission vehicles and engages stakeholders to help demonstrate the viability of low-carbon delivery tools. This kind of policy innovation is precisely what is needed to address the many open questions relating to how cities can reduce their transportation emissions.

5

SCALING UP RENEWABLE ENERGY

Large amounts of renewable energy will be essential to achieving the GHG emission reductions needed to reach cities' climate goals. Federal policy and state governments drive much of the policy to scale up renewable energy generation, but local governments also play a role: more than 180 cities, towns, and counties have pledged to achieve a 100 percent renewable energy supply.[1] Cities can play a significant role in fostering renewable energy at the scale of both utility generation and distributed generation.

The legal authority of municipalities with respect to energy use is often an indirect effect of their status as a consumer and large stakeholder and of their having some tangential regulatory authority (particularly land use authority to catalyze the growth of distributed generation). Most legal parameters that govern municipal options with respect to electricity and energy are set at the state and federal level. However, local governments are (1) market participants with some leverage that comes with large energy purchases; (2) regulators with some peripheral authority over electric utilities, as through land use controls; and (3) large stakeholders in the electricity regulatory process.

UTILITY-SCALE ENERGY GENERATION AND DISTRIBUTED ENERGY GENERATION

It is estimated that global energy demand will rise nearly 50 percent by 2050[2] and that almost all of this energy will need to be carbon free.[3] Both utility-scale and distributed generation are critical, and they can be thought of as complementary:

- *Utility-scale energy generation* brings in the bulk of a city's needed power at the community level. Projects are usually considered utility scale at ten megawatts or larger. The legal constraints on a city in procuring utility-scale renewable energy stem mostly from the state's regulatory regime.[4]
- *Distributed energy generation* yields energy at a smaller scale, such as through rooftop solar panels. In addition to increasing capacity for renewable energy, some sources of distributed energy can also increase resilience by providing power during a utility outage.

Electricity regulation is complicated, with wholesale generation and sales regulated at the federal level by the Federal Energy Regulatory Commission and retail sales regulated by state public utility commissions. Additional complexities include whether a state is traditionally regulated (i.e., utilities have vertically integrated monopolies) or "deregulated" (i.e., the generation and distribution functions of electricity service have been split, and consumers have at least some choice in their supply generator) and whether a state sits in the territory of an independent system operator or a regional transmission organization. Municipalities can advocate for changes to the energy regulatory regime at the state level but mostly have to work within the parameters of existing frameworks.

ELECTRICITY VERSUS ENERGY

While national and global carbon mitigation will require a diverse range of clean energy sources, in the context of a city, decarbonization often focuses on electrifying as much as possible—especially in the building and transportation sectors—and greening the electricity supply. Cities may have some reason to encourage the development of nonelectricity energy sources, but, for many cities, carbon-free electricity will be the main source of energy used to reduce local GHG emissions, particularly as more electric vehicles and electric heat pumps are deployed. Therefore, *electricity* law is particularly relevant in explorations of renewable *energy* policy for cities.

EQUITY AND LOCAL ENERGY POLICY

Decisions about how and where to develop renewable energy resources can give rise to significant questions about equity, energy justice, and reducing co-pollutants. City policies to encourage renewable energy generation, for example, can be inclusive and offer meaningful opportunities for low- and moderate-income residents, and other members of frontline communities and communities of color to partake in the local benefits of phasing out dirty energy sources—or they can further entrench existing inequalities. Moreover, in some places in the United States, energy costs can reach nearly 20 percent of a low-income household's income,[5] and some energy policies threaten to push costs even higher. Finally, environmental justice communities have long been more likely than others to be situated near heavy-polluting power plants,[6] and advocates often push for the closure of dirty plants in low-income neighborhoods and communities of color as new renewable resources come online.

SETTING A RENEWABLE ENERGY GOAL

A local law committing a municipality to a 100 percent renewable energy supply or to net-zero emissions can be a powerful tool to catalyze the development of renewable energy. By setting a goal and measurable interim targets, a municipality can assess its clean energy needs, track progress, and communicate demand to energy suppliers. A goal also provides an opportunity for community engagement and input in the drafting and implementation process and can help city staff marshal internal resources to support the goal.

These local laws take a variety of forms. Many cities have goals for reducing GHG emissions[7] that will necessarily require an increase in the renewable energy supply. The most well-developed of these targets are fleshed out in climate action plans that describe the strategies a city will use to achieve its GHG emission reduction or net-zero goal. Ideally, such a climate action plan would be approved by the city's elected leaders and enacted as law. For example, Ann Arbor, Michigan's A2Zero Carbon Neutrality Plan includes plans for community choice aggregation, on-site renewable energy generation with storage, a community solar project, and a larger-scale solar project.[8] The Ann Arbor City Council approved the A2Zero plan in June 2020.[9]

PROCURING UTILITY-SCALE RENEWABLE ENERGY

To meet their climate goals, cities need to procure enormous quantities of renewable energy. They do so largely subject to state laws and regulations, meaning that the purchasing options for renewable energy available in one state may be precluded in others. One significant difference among state regulatory regimes is whether customers, including local governments, have access to "retail choice" (i.e., the ability to choose from a number of electricity generators to obtain a better rate or a higher content of renewable energy), but there are also many other variations in state energy regulation. Municipally owned utilities and cooperatives may also be subject to alternative regulatory requirements.

Several contractual arrangements are emerging as templates for the municipal procurement of renewable energy. In pursuing any of these, a municipality must be mindful of state and local procurement requirements and guidance with respect to any public contract.

Green Tariffs

As large customers, municipalities may have access to "green tariffs" from their electric utility companies. A green tariff is an offering by a local utility that allows certain customers to procure electricity from sources of renewable energy.[10] Under a green tariff, the utility contracts directly with the renewable energy generator, while the customer simply pays the green tariff rate to the utility for electricity and the associated renewable energy credits.

A green tariff raises relatively few legal barriers for a municipality; because the contracting arrangement goes through the local utility, it does not depend on access to retail choice.[11] However, given that state public utility commissions have authority over electricity rate-making, any green tariff rate structure will need to be approved by the public utility commission, a process over which a city has little direct control but might be able to exercise some leverage given its position as a large customer and stakeholder. In addition, the terms of a green tariff contract can be onerous. They may require lengthy commitments or prohibit a green tariff customer from net metering.[12] Finally, a green tariff may not scale up renewable resources locally, particularly if it relies on faraway renewable power generation.

Power Purchase Agreements

Rather than purchasing renewable energy through a green tariff or other direct agreement with the local utility, cities in jurisdictions that have retail choice can purchase a large quantity of renewable energy through a power purchase agreement (PPA) with a renewable energy project developer. The terms of a PPA usually set a price per unit of energy and specify that the power purchaser takes title to the energy at the delivery point.[13]

A PPA gives rise to meaningful legal obligations. In particular, because the power purchaser (also called the "offtaker," who in this case is the municipality) takes title to the energy, it must be licensed as a power marketer by the Federal Energy Regulatory Commission or hire a third party who is.[14] In addition, the state's energy regulatory framework weighs heavily in a municipality's ability to enter into a PPA. In states with so-called traditionally regulated electricity markets in which consumers do not have access to retail choice, municipal customers are generally unable to obtain renewable power from anyone other than their utility (with some exceptions for municipal utilities and community choice aggregators).[15]

Community Choice Aggregation

Community choice aggregation (CCA) is a program through which an aggregator—often, though not always, a municipality—pools the electricity demand of many or all residents, leveraging group purchasing power to offer an alternative to the electricity generation provided by the local utility.[16] Under CCA programs, the existing utility continues to provide all transmission, distribution, and billing services, while power generation is provided by a third-party power producer.[17]

CCA must be authorized by state law,[18] and there is significant variation among programs. Some CCA programs are focused more on lowering energy costs than on greening the electricity supply.[19] Policy-makers recommend that for maximum impact, a CCA program should be "opt out," meaning that local residents are automatically enrolled unless they expressly indicate that they do not wish to be, rather than "opt in."[20] For example, in the Hudson Valley Community Power program in New York State, the default electricity option for homes and small businesses in fourteen communities is 100 percent renewable energy.[21] Only ten states currently authorize CCA.[22]

CCA can reshape the relationship among the utility, municipality, and electricity consumer, making a city and its residents far less reliant on the local investor-owned utility in scaling up renewable energy. Even exploring a CCA program can yield a heightened commitment to renewables by a utility looking to maintain its monopoly position. At the same time, many utilities enjoy legal protection of their status as a monopoly, and any redrawing of these lines could give rise to questions of applicable law and costs. In California, for example, utilities are entitled to what is known as a power charge indifference adjustment,[23] meant to serve as an "exit fee" paid by CCA customers to compensate for utility investments in electricity infrastructure over many years.[24]

FRANCHISE AGREEMENTS

How might a city get a reluctant utility to the table? There is no one-size-fits-all answer, but some municipalities have in recent years made creative use of their utility franchise agreements as leverage in utility negotiations. Franchise agreements are usually straightforward: the municipality provides a local utility (electric, gas, cable/internet, or otherwise) with access to the public right-of-way, which the utility uses to install the wires, poles, and other infrastructure needed to provide the utility's services to local residents. Often, the utility pays a "franchise fee" to the municipality for this access.[25]

Municipalities do not universally have franchise agreements with their local electric and gas utilities—some states' laws do not allow for them, and where they do, some are decades away from their expiration dates.[26] Further, franchise agreements are limited by state law and sometimes may not govern anything more than the terms of utility access to the public right-of-way. Still, for municipalities that do have franchise agreements, and especially for those with franchise agreements set to expire in the next several years or with the possibility of an early exit, a franchise renegotiation can provide needed motivation for a local utility to consider demands made by a municipality for a higher quantity of renewable energy serving the community, expanded energy efficiency programs, and energy retrofits in underserved neighborhoods. Though franchise agreements are limited in scope, in jurisdictions where they are used a utility often prefers to have one in place to protect their business interests.

A limited number of city–utility franchise agreements contain provisions relating to clean energy, the strongest ones being concentrated in Colorado. For example, several electric franchise agreements in Colorado agree to "invest in clean, renewable electric power and include renewable resource programs,"[27] while others commit "to actively pursue reduction of carbon emissions . . . with a rigorous combination of energy conservation and energy efficiency measures, clean energy measures, and promoting and implementing the use of renewable energy resources on both a distributed and centralized basis."[28] Most other city–utility franchise agreements do not mention renewable energy.

OTHER UTILITY AGREEMENTS

Rather than enshrining clean energy goals in a franchise agreement, some cities and utilities have stand-alone agreements to increase clean energy supply and improve energy efficiency offerings. Some, like Minneapolis and Salt Lake City, leveraged their franchise negotiations to reach such agreements. Minneapolis entered into a "Community Energy Partnership" with its electric and gas utility companies addressing energy efficiency, renewable energy generation, and community outreach.[29] The Community Energy Partnership has worked iteratively to develop programs to reduce GHG emissions, including strategies to eventually achieve a 100 percent renewable electricity supply.[30] Similarly, Salt Lake City entered into a "Joint Clean Energy Cooperation Statement" with its utility that covered, among other matters, a three-megawatt sale of solar energy to the city for its municipal facilities.[31] (Unfortunately, the agreement also prohibited the city from implementing CCA during its term.)

Outside the franchise agreement process, the city of Charlotte, North Carolina, entered into a memorandum of understanding to a "establish a low carbon, smart city collaboration" with its electric utility.[32] While specific targets are left to a to-be-developed work plan, and while renewable energy is called out only in passing,[33] such a memorandum of understanding could be used by a municipality and a utility to agree more expressly to renewable energy goals at both the utility and distributed scales. Interestingly, Charlotte's memorandum of understanding acknowledges that the utility's actions to support the partnership "may be subject to North Carolina regulatory utility requirements" and states that the parties agree to cooperate on regulatory approvals if needed.[34] Sarasota, Florida; Madison, Wisconsin; and Denver, Colorado, also have city–utility partnership agreements.[35]

MUNICIPALLY OWNED UTILITIES

Many municipalities have owned local electric or natural gas utilities for decades. More recently, the municipalization (the process of transferring to local government ownership) of energy utilities has gotten another look in the effort to scale up renewable energy. Historically, municipal utilities have not focused on environmental metrics, and the evidence that public utilities achieve higher supplies of renewable energy than private investor-owned utilities is limited. In some states, municipal utilities have even been able to avoid complying with renewable energy requirements applicable to investor-owned utilities.[36]

Still, some municipal utilities make electricity decarbonization a priority. For example, Austin Energy in Texas offers robust renewable energy options to consumers, including a 100 percent wind power option, incentives for rooftop solar panel installations, and access to community solar programs.[37] The utility currently claims that about one-third of its electricity is renewable and that it is on a pathway to 55 percent renewable electricity by 2025.[38] The public utility in Jefferson County, Washington, boasts a 97 percent carbon-free energy supply (most of which is hydropower, a resource purchased from the federally owned Bonneville Power Administration,[39] which prioritizes sales to publicly owned utilities).[40] Municipalities that already have public electric utilities can direct them to procure additional renewable energy or, more directly, enforce a local renewable portfolio standard or goal. (Early efforts to explore decarbonization for municipally-owned natural gas utilities have been met with significant resistance.)

For cities with nonmunicipal, investor-owned utilities (IOUs), the decision whether to municipalize the electric utility is a complicated one, legally and financially. While nearly all states allow electric utility municipalization, the details vary.[41] Municipalities may generally purchase the local IOU and in some states may acquire it by condemnation.[42] The price to acquire the IOU is informed by state law or a franchise agreement's pricing terms.[43] Just compensation for condemned utility assets would also be determined by state law applicable to eminent domain.[44] Additional costs can include "stranded costs": compensation for investments made by an IOU in reliance on the ability to sell power into a market.[45] Significant litigation can arise from all of this.

In addition, some state laws require municipalization to be approved by voters,[46] a process that can not only throw a wrench in municipalization plans but also gives the local IOU a lengthy opportunity to conduct a potentially well-funded public campaign to keep electricity service private.[47] Several other legal considerations need to be taken into account as well, including whether state public utility commission approval is needed[48] and whether the municipality has the authority needed to finance the purchase of the IOU.[49]

DISTRIBUTED RENEWABLE ENERGY

In many ways, distributed energy—a category that includes not only generation resources like rooftop solar panels but also microgrid and storage assets—plays out in the shadow of state public utility law, just as utility-scale transmission does. State law may set requirements or limitations around who may own, distribute, or sell energy from these distributed systems, and state public utility commissions generally set the terms and pricing for net metering, the process by which the owner of a distributed energy source sells electricity back to the grid. But cities can also take affirmative steps to incentivize and shape the development of distributed energy. This section discusses legal tools that can facilitate distributed generation, as well as legal issues that can arise with an expanded system of distributed renewable energy.

Distributed Solar Energy

Perhaps the most visible resources of distributed generation are the rooftop solar panels installed on a wide array of residential, commercial, and industrial buildings. Rooftop solar panels could satisfy as much as 39 percent of the energy needs of the United States,[50] and research shows that homeowners are more likely to install them when they see their neighbors invest in the technology.[51] City policies that facilitate the installation of solar panels can therefore help a community develop renewable generating capacity relatively quickly and with little investment by the city itself.

Solar Zoning

Municipal zoning laws can either hinder or support the use of rooftop solar panels. Zoning codes are often unclear about the treatment of solar panels, making solar energy a less appealing investment for building owners unsure about whether on-site solar power generation is permitted in a particular zone. However, zoning codes can in some jurisdictions require and in others offer incentives for development that includes rooftop solar panels.

Simple tweaks to a local zoning code can help eliminate the uncertainty that unclear zoning codes create. For example, the city of Aurora, Colorado, passed a local ordinance in 2011 to permit solar and wind energy systems in all districts (subject to all other building and zoning codes; for example, solar energy systems must meet zoning height requirements).[52] Austin, Minnesota, similarly clarified that "solar energy systems are allowed as an accessory use in all zoning classifications" (again, subject to height, setback, and other zoning requirements).[53] These kinds of zoning modifications have become increasingly widespread in recent years.

A larger number of tools are available to affirmatively promote or require solar energy development. Cities across the country have enacted a variety of green zoning requirements (some of which are discussed in chapter 3). In some cities, a delineated geographic zone is subject to green building or energy requirements, while in other cities, incentives for green building apply to a whole community. In the city of Watertown, Massachusetts, a zoning ordinance requires buildings of ten thousand square feet or more or with ten or more residential units to include a solar energy system.[54]

Solar Building Codes

Local laws can authorize or mandate the use of rooftop solar panels. New York City, for example, amended its building code in 2019 to require new buildings and buildings undergoing major roof renovations to cover the roof with solar panels, a green roof system, or a combination of the two (exceptions are granted where solar panels and green roofs are infeasible and to allow some limited space to be used for a terrace or other recreation area).[55] Many cities in California also have solar energy requirements[56] (building on California's state-level requirement that many residential buildings include solar installations),[57] some of which complement all-electric construction efforts. For cities not ready to require solar panel installations, a "solar-ready" building requirement can help facilitate solar panel installations when the technology becomes economically feasible. Seattle, Washington, has both a small-scale solar energy requirement and a larger-scale solar-readiness requirement in its building code for commercial buildings.[58] Tucson, Arizona, has had a solar-readiness requirement for single-family and duplex housing since 2009.[59] The Massachusetts Stretch Energy Code, which may be adopted by municipalities in the state, also includes a solar-readiness requirement for new residential buildings.[60]

In addition to these direct mandates, solar energy can be used as a compliance pathway for a building code requirement, or it can be made implicit in a more general mandate. For example, Boulder, Colorado, requires new construction to achieve a maximum "Energy Rating Index" score; on-site renewable energy systems can factor into this score, encouraging builders to include solar technology.[61] Mandates or incentives for net-zero or passive-house buildings also compel the expansion of distributed solar energy as builders look to renewable energy as a pathway to comply with net-zero requirements.

Finally, building codes can require that certain energy-intensive systems be powered by renewable energy. Boulder's energy conservation code requires any energy use for residential pools, spas, and snow-melt systems to be offset by on-site renewable generation.[62] Hot-water heaters are also heavy users of energy and can be good candidates for solar energy use; thus, cities might consider requiring that new buildings use solar-powered hot-water heaters, depending on the local climate.[63]

Many cities lack the authority to set or amend building codes locally. These cities would need to rely on local zoning authority or consider passing a solar energy requirement through a stand-alone ordinance (subject to challenges that such a requirement would be preempted by a state building code or other state law).

Streamlining Solar Permitting Requirements

Estimates have put the cost of local regulatory requirements for rooftop solar panel installations at upward of $1,000 per system.[64] While some experts favor statewide permitting standards for solar energy that are consistent from city to city (or even eliminating rooftop solar permitting requirements entirely),[65] in the absence of state regulation municipalities can accelerate the development of solar power by streamlining local permitting requirements. One idea is to enact a "uniform solar permit" within a municipality that simplifies the approval process; for example, municipalities in New York State's Clean Energy Communities Program are encouraged to adopt a provided model uniform permit.[66] Cities might also waive permitting fees (as in South Miami, Florida)[67] or expedite permitting reviews for residential solar panel installations (as in Boise, Idaho,[68] and Los Angeles and San Jose, California).[69] Finally, cities can help facilitate a smooth permitting process by making information about all requirements available in one place online.[70]

Community Solar

Community solar—an array of solar panels at an off-site location that multiple customers sign onto to increase the use of local solar energy—is a way to give access to solar power generation to those who cannot install their own rooftop solar panels because they rent their homes or businesses, live or work in multi-unit buildings, or have limited access to sunlight. The contractual specifics of community solar arrangements vary, but they generally follow either ownership models, in which a number of people own panels at the project site (which might allow owners to take advantage of tax credits) or subscription plans, in which customers subscribe to a community solar project for some period of time.[71] Owners or subscribers are credited for their share of the solar energy generated at the project site.[72] Community solar can help ensure equitable access to renewable power: community residents who do not own their homes or are unable to shoulder the up-front costs of solar panel installation can reduce their GHG emissions and their electricity bills by subscribing to a community solar project. Ideally, community solar projects that serve environmental justice and frontline communities are located in those communities to help reduce localized air pollution from fossil fuel power sources,[73] though the community solar framework does not require this.

A number of legal issues can arise in connection with community solar projects, though they are usually not gating items. With careful structuring, community solar projects can generally be designed to comply with applicable laws. The large number of projects built to date (according to the Solar Energy Industries Association, community solar projects generating more than five gigawatts of energy have been built across forty-one states[74]) suggests that most of the potential legal considerations have been worked through in many jurisdictions. Many states have laws or utility regulations relating to community solar projects, requirements that may include caps on capacity for individual projects or for the state's entire community solar program, customer rates (often set with the input of state public utility regulators, who have rate-setting authority), subscriber requirements (e.g., a minimum number of subscribers or a preference for low- and moderate-income customers), and geographic restrictions (some proximity to subscribers may be required).[75]

Cities can help facilitate community solar projects by clarifying zoning, siting, and permitting requirements and other rules applicable to small-scale solar projects. Cities might consider making the permitting process as simple as reasonably possible for parcels that are ideal for solar energy and that comply with any applicable geographic restrictions imposed by state law. Cities can also simplify the permitting process for community solar projects located in environmental justice areas or that offer subscriptions to low- and moderate-income customers.

RENEWABLE ENERGY FOR MUNICIPAL BUILDINGS AND OPERATIONS

As with other policies, a city might begin requiring solar energy development by starting with its own buildings. Such an activity avoids many of the questions about municipal authority to enact building or zoning requirements and can help demonstrate leadership in expanding renewable energy. Determining the right contracting arrangement for solar panels on municipal sites will require the input of the city's attorney, and options will be informed by the state's electricity regulatory framework. For example, cities will need to determine whether to enter into a PPA to purchase the electricity generated by privately owned solar panels installed on municipal property (as Washington, DC, did), lease the solar panels (the structure chosen by Kansas City, Missouri), or choose another contracting arrangement.[76]

ENERGY STORAGE

Energy storage is a key complement to distributed solar energy generation, and many of the land use and building regulation tools applicable in the context of solar energy can be adapted to facilitate increased energy storage at the distributed scale. Solar-plus-storage systems, as they are called, can help customers with their demand for energy from the grid by storing energy generated during peak sunlight hours for use later in the day and can offer backup electricity in case of a utility outage. These features can improve the economics of a building solar system.[77]

Mitigating demand peaks by fulfilling those peaks with stored energy helps lessen the overall supply of renewable energy that needs to be built up to meet a community's needs. Cities can make use of the same types of tools to clarify zoning and permitting requirements for energy storage that they use for solar energy siting. In particular, cities could ensure that rooftop solar-plus-storage systems are treated with no more burdensome requirements than solar-only installations.[78] Cities can also encourage the development of energy storage by supporting the development of microgrids (discussed shortly).

Vehicle-to-Grid Technology

A novel approach to scaling up the storage of distributed energy has emerged among local governments considering using school bus batteries to store energy while they are idle.[79] While vehicle-to-grid (V2G) technology is relatively new, school buses represent a promising opportunity for V2G because they sit unused in the middle of the day and during the summer, two periods of peak energy demand.[80] Some local governments enable this technology by removing legal or permitting obstacles to charging and V2G infrastructure for school buses and by piloting electric school buses on their own or in collaboration with their local utilities. In addition to providing energy storage and reducing transportation GHG emissions, electrifying school buses has the added equity benefit of reducing exposure to air pollution among children and in low-income communities.[81]

MICROGRIDS

Microgrids, which are local energy grids that can disconnect from the traditional grid and operate autonomously, sit squarely at the intersection of climate change mitigation and adaptation.

While they are often hailed as a tool to enhance resilience (as they can be cleaved from the grid during storms or other outages and continue delivering power internally),[82] they can also help mitigate GHG emissions with the use of solar panels, energy storage, and demand-response techniques, the latter two of which reduce the need for carbon-intensive peaker plants, which are generally turned on only during periods of high demand.[83] They also transmit power more efficiently than utility-scale sources, as the energy travels a shorter distance from the source of generation to the end user.[84] Legal tools for municipalities to support microgrid development include (1) waiving permitting fees and expediting permitting, (2) granting zoning incentives and revising zoning codes that hinder the on-site generation and storage of renewable energy, (3) reviewing franchise agreements for potential conflicts with microgrid generation, (4) establishing "district energy zones that provide municipal infrastructure that will allow future microgrid development," and (5) supporting project development by identifying ideal locations and providing information about underground infrastructure.[85] Mixed-use zoning can also support microgrid or "energy district" infrastructure, as energy demand varies by building use and time of day.[86]

DISTRIBUTED WIND ENERGY GENERATION

Distributed wind energy generation is relatively less common than distributed solar and can give rise to legal challenges in urban and suburban areas. Some legal considerations are similar to those relating to rooftop solar: zoning requirements can help or hinder wind energy development. In another similarity to solar development, wind energy projects can be stymied by a lack of wind access rights or a lack of clarity about what access rights exist. Local laws can help ensure some protection for wind access by protecting wind channels from neighboring projects through "air easements" or "air access zoning regimes" (similar to the Cambridge, Massachusetts, program that allows developers to register solar projects and receive some limited protection of their solar access rights).[87]

In one significant difference from distributed solar energy, wind turbines move by design. They can be noisy and generate flickers of shadow and light.[88] These features can give rise to nuisance claims, and cities may play some role in delineating how both wind energy development and its potential impacts on neighbors are treated. Local governments can require setbacks, height limits, and noise (decibel) limits, as they often do for oil and gas development.[89] Zoning, which can make clear where and which type of renewable energy development is permissible, is a common legal tool used to enact these restrictions. Unfortunately, some state and local laws hinder wind development through outright bans (for example, Kansas allows local governments to prohibit wind energy development)[90] or overly onerous environmental or permitting requirements.[91]

THE SMART GRID, SMART METERS, AND OTHER SOURCES OF INFORMATION

Distributed energy generation can be inhibited by a lack of information about where such resources should be placed or whether they would decrease the carbon intensity of—or costs associated with—community energy use. Technology associated with the so-called smart grid, "a general term to describe the computerization of both electricity transmission and distribution wires and certain appliances attached to the grid,"[92] can be used to better direct energy resources where they are needed based on real-time data, helping the grid be more energy efficient and reliable, manage demand load, and, if desired, prioritize renewable energy.[93]

Local legal requirements around smart grid technologies, particularly smart meters (which track building or tenant/unit energy use more or less in real time), can help inform energy supply and use decisions, including the decision to install solar panels or another renewable resource. Smart metering and other energy disclosure requirements can give rise to legal concerns regarding privacy. For example, the U.S. Court of Appeals for the Seventh Circuit held smart meter data collected every fifteen minutes to be a "search" under the Fourth Amendment of the U.S. Constitution (and therefore a protectable privacy interest) but that in Naperville, Illinois, "significant government interests" (i.e., "the modernization of the electrical grid") in using such data made the search reasonable.[94] Smart grid technologies, with their access to large troves of data, can also implicate data security and cybersecurity considerations, which cities must be mindful of in developing smart grid or smart meter programs.[95]

6

DECARBONIZING A CITY'S WASTE

Of the sectors accounted for in the *Global Protocol for Community-Scale Greenhouse Gas Emission Inventories*,[1] a city's waste stream generally has the smallest carbon footprint. Many U.S. cities' waste accounts for around 5 percent of their GHG emission inventories.

Waste-related GHG emissions are deeply intertwined with the carbon output of other city sectors, especially transportation, as well as with global patterns of production and consumption. Waste that ends up in a city's trash usually began its life cycle somewhere else; many cities produce relatively few products, and no city can produce everything its residents use. Thus, a city's efforts to rein in waste-related GHG emissions focus on the relatively limited portion of an item's life cycle during which it passes from the consumer to a waste bin to a processing facility, and finally to a landfill or to continued life as a recycled product. This chapter focuses on these end stages of a widget's life cycle. It concludes by discussing the broader context of global production and consumption and the movement toward a circular economy.

THE BASICS OF GREENHOUSE GAS WASTE ACCOUNTING

Cities calculate the GHG emissions associated with their municipal solid waste, sludge, industrial waste, wastewater, clinical waste, and hazardous waste. The emissions attributable to these waste streams include emissions from waste processing through aerobic or anaerobic decomposition, incineration, and waste decomposition in a landfill. A city's waste-related

A NOTE ON EQUITY

Waste hauling, processing, and landfilling are sources of significant historical and ongoing damage to environmental justice communities. Waste-processing facilities, landfills, and trash-incineration sites are overwhelmingly more likely to be sited in low-income communities and communities of color than in more prosperous and whiter neighborhoods, and the high rates of childhood asthma and other diseases attributable to local air pollution reflect this injustice. Policy aimed at reducing waste-related GHG emissions overlaps with—and either advances or inhibits—equity-related objectives in four main ways. First, environmental justice communities are often disproportionately burdened by the siting of new waste-processing facilities. Second, such communities are often located near waste depots and therefore experience significant traffic and local air pollution from waste-hauling trucks. Third, lower-income communities may lack zero-waste infrastructure like composting programs at local parks and community gardens. These programs can provide co-benefits to nearby residents by offering open space, garden mulch, and opportunities for community engagement. Fourth, low-income communities may be unable to bear the financial burden of waste decarbonization programs like fees for paper bags or a "pay as you throw" program, necessitating the need for policy solutions to mitigate these costs.

GHG emissions include those from waste that originates in the city, *wherever disposed of*; unlike other GHG sectors, waste GHG measurements do not stop at the city boundary. GHG emissions relating to waste processing but attributable to other sectors, like stationary energy and transportation, are counted in those other sectors.

BASIC LEGAL FRAMEWORKS

A city's waste reduction efforts play out against the backdrop of three legal frameworks: (1) federal law and the Dormant Commerce Clause, (2) state law, and (3) contract law.

The Dormant Commerce Clause and *Philadelphia v. New Jersey*

Because of the Dormant Commerce Clause's restrictions on local laws that discriminate on out-of-state economic actors, local governments generally may not enact waste policies that favor private local waste facilities (see chapter 1.) In *Philadelphia v. New Jersey*, the U.S. Supreme Court struck down a New Jersey law banning the importation of nearly all "solid or liquid waste which originated or was collected outside the territorial limits of the State" as in violation of the Dormant Commerce Clause.[2] Subsequently, in *C&A Carbone, Inc. v. Town of Clarkstown, N.Y.*, the court struck down a local flow control ordinance requiring all waste handled in town to go through a single privately owned transfer station, noting that "[d]iscrimination against interstate commerce in favor of local business or investment is *per se* invalid" unless the government can show that it had no other means of achieving its legitimate goals.[3] The court also held that the town's identified purpose of generating revenue to support the transfer station was not a "legitimate local interest" that justified the discriminatory treatment.

Cities have more latitude to enact policies that favor *publicly* owned waste facilities. *United Haulers Association, Inc. v. Oneida-Herkimer Solid Waste Management Authority* clarified the earlier decisions, noting that they did not extend to publicly owned waste facilities because "disposing of trash has been a traditional government activity for years, and laws that favor the government in such areas—but treat every private business, whether in-state or out-of-state, exactly the same—do not discriminate against interstate commerce for purposes of the Commerce Clause."[4] In other words, a local government *may* require that local waste be sent to a publicly owned and operated facility, particularly if balanced with local benefits such as "enhanced incentives for recycling" and local "ability to enforce recycling laws."[5]

Consequently, local governments are barred by the Dormant Commerce Clause from enacting most restrictions that treat out-of-town or out-of-state waste less favorably than local waste, even if they advance the permissible goal of reducing waste-related GHG emissions.

State Law

Solid waste collection and processing, often regulated and managed at the local level, is nested within state law. State laws relating to waste collection can preempt conflicting local laws, but most state legislatures have not regulated so expansively as to preempt *all* local controls. For example, many local jurisdictions have zoning controls in place relating to waste-processing facilities; as long as these controls do not conflict with state law, they are generally permissible exercises of local authority.

Take Connecticut, for example: the state's Department of Energy and Environmental Protection (DEEP) oversees the permitting and operations of waste disposal sites, but local governments may regulate solid waste facilities through zoning requirements or other local laws.[6] A separate state agency operates certain solid waste management facilities in the state, and local laws can be preempted if they conflict with a permit granted by DEEP for these facilities.[7] Or Pennsylvania: the state's Solid Waste Management Act sets many of the requirements relating to the operation of solid waste facilities.[8] Local governments may enact zoning requirements as long as they are not inconsistent with the act or with the state's Department of Environmental Protection requirements. The details can get quite granular: municipalities may not restrict the days or hours that waste facilities operate,[9] but they can set limitations relating to when waste is trucked in and out.[10] In Georgia, a state authority also has responsibility for permitting solid waste disposal, though the applicant must certify compliance with both local zoning requirements and any local or regional solid waste management plan to obtain a permit.[11]

In all states, whether zoning requirements apply to governmental entities depends on the facts and relevant law. While many local governments have significant authority, cities could be impeded by state law in siting new recycling, anaerobic digestion, or waste-to-energy facilities, or in limiting or altering the operations of landfills, tipping floors, or transfer stations.

City Legal and Contractual Obligations

Municipalities employ a handful of legal and contractual tools to guarantee an agreed financial return to waste-processing and disposal facilities. These tools, which are generally legal obligations of a city with potential consequences under contract, can (1) require a municipality to provide a larger quantity of nonrecyclable waste than is optimal given their waste reduction goals or, alternatively, (2) be used creatively to incentivize increased waste diversion.

The first of these resources is a *flow control law*. Flow controls—the requirements at issue in both *C&A Carbone* and *United Haulers Association*—are legal requirements that a government cause a minimum quantity of waste to be delivered to a landfill or waste facility.[12] Flow controls are generally used to ensure that a waste facility will be profitable or able to repay its bonds. Relatedly, a *"put or pay" contract* sets a minimum amount of waste disposal that a government must pay a facility for, regardless of the actual amount of waste delivered.[13] A municipality may also be a party to a *landfill gas contract*, which could guarantee an amount of waste or money in the exchange of waste for energy.[14] Finally, depending on the legal authority delegated by state law, municipalities can assess *taxes or fees on waste* (e.g., nonrecyclable, recyclable, organic), whether it be tipped, dumped, collected, or transported.

All these mechanisms affect the incentives for waste diversion and reduction. They can entrench practices that do not align with a city's waste reduction and diversion goals and may put meaningful practical limitations on zero-waste goals if they are not revisited. At the same time, versions of these legal and contractual tools could be leveraged to increase investment in desired waste-processing facilities such as composting and recycling facilities or anaerobic digesters. Arrangements like flow controls can also serve equity ends: they can ensure the use of designated waste-hauling routes or processing locations to reduce local air pollution in environmental justice communities or reduce GHG emissions from waste transport.

MAKING THE COMMITMENT: ZERO-WASTE GOALS AND PLANS

Setting a goal is often an early step for cities looking to reduce waste and associated carbon emissions. These goals can take the form of a stand-alone zero-waste goal or be embedded in broader carbon reduction goals. Some more aggressive cities codify their waste diversion goals in law.[15] Zero-waste goals in and of themselves are subject to relatively little legal scrutiny, which is more likely to arise in response to the actions undertaken by a city to carry out its zero-waste mandate.

LEGAL TOOLS FOR REDUCING WASTE AND WASTE-RELATED GREENHOUSE GAS EMISSIONS

Two broad concepts are essential to reducing waste stream–related GHG emissions, particularly the methane released as waste breaks down in a landfill. The first is *source reduction*. Policy tools like bans on single-use plastics and "pay as you throw" are aimed at source reduction. The second is *waste diversion*. Mandatory organics recycling, increased composting, and scaling up recycling capacity are ways to advance waste diversion efforts. There is some overlap across categories—a pay-as-you-throw policy, for example, can also help divert waste from a landfill if it prices trash higher than recyclables or organic waste.

This chapter discusses legal issues confronting policy tools for waste reduction and waste diversion. These policy tools include (1) materials-related product bans, (2) pricing trash ("pay as you throw"), (3) organics recycling, (4) small-scale composting, (5) the siting of waste and recycling facilities, and (6) reforms to waste handling and transport.

MATERIALS-RELATED PRODUCT BANS

An increasing number of cities (and states) have enacted bans and restrictions on products based on the materials they are made of: single-use plastic shopping bags, containers made from expanded polystyrene (EPS) foam (commonly known by the brand name Styrofoam), and, less commonly, other single-use plastic items. Generally, these products are not banned outright but rather are restricted as free handouts at a point-of-sale location such as a grocery store or take-out restaurant. To date, these actions have been largely directed at decreasing the number of these items that enter a city's waste stream or preventing nuisances like plastic bags getting stuck in trees, but product bans can impact GHG emissions in significant ways; perhaps most importantly, these bans, once proven successful, may be able to target carbon-intensive products, based on either direct emissions or life-cycle emissions.

Plastic Bags

Plastic bag bans—prohibitions on handing out free single-use plastic bags at points of sale—have become so prevalent in U.S. cities and states that a backlash has developed in the form of both state preemption and misinformation campaigns. Like other climate-benefiting laws and policies, plastic bag bans often reduce waste-related GHG emissions in an indirect way, with the policies framed first and foremost as measures to reduce litter.[16]

State law informs local policy on plastic bag bans and may preempt a ban altogether. State law expressly preempts municipalities from enacting local plastic bag bans in Arizona, Colorado, Idaho, Iowa, Michigan, Minnesota, Mississippi, Missouri, North Dakota, Oklahoma, South Dakota, Tennessee, Texas, and Wisconsin.[17] Even where not preempted, the legal landscape with respect to plastic bag bans varies greatly by state. Some states have statewide plastic bag bans, freeing local governments from having to consider enacting their own bans but potentially locking them into undesirable policy features. In other states, some local governments have passed bans, and, in still others, like New York, municipalities are required by state law to pass a plastic bag ban in accordance with parameters established by the state legislature.[18] In Hawaii, local bans across the state's more populous counties amount to a "de facto statewide ban."[19]

In cities where the state has neither enacted a statewide plastic bag ban nor preempted a local ban, state law can inform policy choices. These choices include whether to impose a fee for paper bags (and if so, how much) and, where a fee is charged, how to use the revenue.[20] While local governments generally have some authority to impose fees, this authority may come with limitations.

Local plastic bag bans have been the subject of significant litigation. Aside from disputes relating to state preemption, there are several categories of claims.[21] First, some have argued that plastic bag bans may yield an increase in the use of paper bags, leading to underexamined environmental outcomes; this argument has not met with much success.[22] Second, some have argued that a fee for bags amounts to an impermissible tax. Many cities do not have authority to impose new taxes without state approval, so—depending on the particulars of state law that distinguish taxes from fees—bag charges for which funds collected go to the government could be held to be impermissible taxes (see chapter 1).[23] Third is a catch-all category that includes complaints by retailers, plastic bag producers, and others required to comply with the terms of a plastic bag ban or who have an interest in seeing it struck down. In New York State, for example, a group of grocery store and bodega owners filed suit alleging inconsistencies between the state's bag reduction[24] and bag recycling[25] laws, unconstitutional vagueness, violation of "anti-gift" clauses of the state constitution because some of the revenue from the five-cent fee would go toward purchasing nonplastic reusable bags for community distribution, and "arbitrary and capricious" regulations that impose "excessive" requirements on plastic bags.[26] The state court agreed with one of the petitioners' claims, striking down a small part of the law but largely leaving the ban in place.[27]

Expanded Polystyrene Foam

Some states and cities ban the sale or distribution of EPS. These bans are generally enforced at points of sale like restaurant takeaway counters. As with plastic bag bans, EPS restrictions have nonclimate benefits like reducing litter and preventing chemicals from leaching into surface water and groundwater; they also can help reduce a city's GHG emissions by keeping trash out of the waste stream.

EPS bans encounter many of the same legal issues as plastic bag bans. They can be preempted by state law, as in Arizona, where a state law preempts local laws that regulate "auxiliary containers," defined as including both "disposable bags" and "extruded polystyrene,"[28] and Colorado, where state law preempts any local law "prohibit[ing] the use or sale of specific types of plastic materials."[29] And they can draw multifaceted litigation from industry players like restaurants and EPS manufacturers. The California Restaurant Association, for instance, sued San Diego to keep its local EPS ban from going into effect, arguing that the city had failed to complete an environmental impact assessment as required by the California Environmental Quality Act and that substitutes for polystyrene would actually produce more GHG emissions than EPS products.[30] Under a 2019 settlement, San Diego agreed to conduct an environmental impact assessment before enforcing an EPS ban.[31] New York City, too, was sued by EPS manufacturing, recycling, and restaurant interests. While the plaintiffs won the first round,[32] a legislative re-do by the city survived legal challenge,[33] and an EPS ban is now in effect in New York City.

Other Single-Use Plastics

It is likely that bans on new products will yield novel legal issues. For example, bans on plastic straws[34] have given rise to questions about whether they run afoul of the Americans with Disabilities Act,[35] as some people with disabilities require a straw to drink. To address these concerns, many cities' straw restrictions either allow plastic straws to be provided upon request or have exceptions for customers with disabilities.[36]

PRICING TRASH: "PAY AS YOU THROW"

Requiring residents and businesses to pay for trash hauling is a common way for cities to reduce waste or at least hold it in check. Alternatively called "save as you throw," the model has been implemented in various ways. In Seoul, South Korea, for example, residents are required to separate food waste from their household trash and must pay to dump food waste—either up-front by purchasing standardized biodegradable bags or by weight into a central collection bin. In the United States, some municipal and private waste haulers require residents to use provided trash bins and pay extra for waste that doesn't fit inside them. Some cities offer multiple bin size options, giving residents meaningful opportunities to save money by producing smaller quantities of waste.

The legal issues with "pay as you throw" are generally light but may involve local authority to assess taxes and fees. For cities with private waste haulers, those waste haulers are generally free to implement a pay-as-you-throw model pursuant to the contracts they enter into with their customers. Municipally owned or operated waste haulers may need to be more careful in developing a fee schedule for residential waste to avoid a judicial determination that the fee amounts to an illegal tax (see chapter 1).

MANDATORY RECYCLING AND ORGANICS RECYCLING

Mandatory recycling for paper, plastic, glass, and metal is well established, if not universal, in the United States. These recycling programs are relatively uncomplicated from a legal perspective, though they theoretically can give rise to the same types of legal enforcement questions as requirements for organics recycling, discussed here.

An emerging tool for increasing waste diversion is the mandatory organics recycling requirement or, alternatively, an organic waste disposal ban. While similar in substance and intent, a mandatory organics recycling law specifies how organic waste must be handled and disposed of, whereas an organic waste disposal ban simply prohibits the landfilling of such waste, leaving generators to figure out what to do with it.[37] A number of states and local jurisdictions have enacted restrictions on organic waste disposal.[38] California, Connecticut, Massachusetts, New York, Rhode Island, and Vermont all have statewide organics diversion requirements, while local organics requirements include those in Austin, Texas; Boulder, Colorado; Hennepin County, Minnesota; the metropolitan Portland, Oregon, area; New York, New York; San Francisco, California; and Seattle, Washington.[39] These restrictions vary in terms of which waste-generating entities are covered (based on weight or volume of waste generated, size of facility, or categorization as a commercial establishment; for example, in many of these regimes, only large restaurants and institutional food-waste generators are covered) and what is required of waste generators, but they all generally aim to reduce organic waste and the associated GHG emissions.

Enforcing mandatory organics recycling laws can give rise to legal questions. First, enforcement mechanisms invite questions about agency authority; for example, is the agency that enforces the ban or diversion requirement (often an environmental agency) the same one that generally oversees restaurants? Second, enforcement against residential generators (i.e., individual households) can raise privacy concerns. A Washington state court struck down provisions of a Seattle law that allowed for a "visual inspection" of private garbage cans as impermissible "warrantless searches" in contravention of the Washington State Constitution.[40] The court otherwise upheld Seattle's ban on organic waste in residential garbage, noting that the city "has the police power to take steps to reduce the quantity of compostable material disposed in landfills."[41] Some cities have nominally mandatory programs but do not actively monitor compliance.[42] Others make organics collection programs voluntary, as New York City did in a multiyear pilot program.[43] A nonmandatory program can avoid legal and policy questions regarding enforcement but may be less effective than mandatory programs; New York's organics program was slow to expand owing to its cost (the voluntary nature of New York's curbside organics collection meant that collection trucks were operating at far below capacity, making for a high cost-to-weight ratio) and was put on hold during the COVID-19 pandemic.

Food Donation

Organic waste requirements necessarily involve decisions about what to do with unused food. When thinking holistically about food waste and food insecurity, organic waste requirements offer cities the opportunity to respond to dual needs and can help make climate and waste policy a key tool in advancing equity objectives. Some local laws take this notion into account by requiring or incentivizing generators subject to diversion requirements to donate food. For example, Hennepin County, Minnesota's organic waste requirement specifies food donation as the "highest priority use" of unused food (though it requires only waste diversion).[44] Austin, Texas, offers credits to generators who donate food that can be used to offset other requirements of the city's waste diversion law.[45] And California has set a target of donating 20 percent of food waste by 2025.[46]

COMPOSTING FACILITIES

Composting facilities large and small can also incur legal scrutiny. As with other forms of municipal solid waste operations, composting operations may be subject to overlapping state and local laws. For example, Texas has permitting, registration, and notification requirements for composting operations, though many operations may be exempt based on their type or the materials they process.[47] Some California air districts have rules on, among other matters, emissions from ammonia, volatile organic compounds, and particulate matter at composting facilities.[48] Other states set material composition standards and regulate the land application of compost.[49] Local government composting operations must comply with state law and assert jurisdiction only to the extent authorized or not preempted by state law.

Small-scale composting operations—at community gardens, small farms, or private homes—face unique legal barriers that can often be resolved with careful planning by local lawmakers but nonetheless lead to uncertainty and underinvestment. The first of these barriers comes in the form of *zoning requirements*, which may prohibit composting facilities in agricultural, commercial, or residential zones.[50] Local governments can avoid this prohibition by allowing composting operations below a certain size or quantity of waste or as an accessory use. The second potential legal obstacle is the *exclusive waste-carting franchise agreement*.[51] While waste-carting franchises or "commercial waste zone" systems are recognized for their GHG emission–reducing and other benefits, exclusive franchise agreements with a large company can leave doubt about the permissibility of smaller-scale composting. Some contracts or franchise agreements can avoid this consequence by including a carve-out for small-scale or community composting.

THE SITING OF WASTE AND RECYCLING FACILITIES

Achieving a city's zero-waste goal—or even increasing waste diversion from landfills—requires facilities that can process or recycle materials, and cities will encounter both opportunities and hurdles in the land use laws that apply to these kinds of facilities. A city's zoning code and permitting processes can facilitate or inhibit the development of recycling and organics-processing sites. Several cities have revised their zoning codes to encourage these facilities. Fresno, California, for example, permits both large and small recycling facilities—large processing plants in certain districts only, with a minimum size of three acres and located far from residential areas, and smaller recycling centers within other businesses as a conditional use.[52] In Madison, Wisconsin, large recycling centers are permitted in identified industrial districts and conditionally permitted in employment districts.[53] Cities may also revise their zoning codes to permit small organics-processing facilities and other needed waste infrastructure.

Waste needs will evolve, and municipalities will need waste-processing sites that are adaptable to changes in recycling needs over time.[54] For example, waste diversion goals, plans, and requirements will necessarily move a city's waste system away from landfilling while increasing demand for reuse and recycling. While a city may not have direct control over the design of a facility unless it owns or operates it, it may be able to use contract terms, financial incentives, or building or zoning code requirements to ensure waste-processing facilities can be adapted to accommodate changing waste needs.

WASTE TRANSPORT EMISSIONS

A city's waste collection and processing systems rely heavily on a multilayered network of vehicles tasked with removing waste from the curb and transporting it to its ultimate destination. Reforms to waste handling and transport can reduce GHG emissions from the transportation sector—a benefit to a city's overarching goal of reducing GHG emissions. Moreover, deploying greener waste-hauling vehicles and practices can, in many cities, help alleviate environmental justice and equity burdens. Replacing fossil fuel–powered garbage trucks with electric ones reduces not only GHG emissions but also local air pollution in the communities through which the trucks drive. Limiting the vehicle miles traveled by refuse-collection vehicles does the same. With landfills and waste-processing facilities often situated in low-income neighborhoods and communities of color, limiting trips through and near those underserved areas and transitioning still-needed trips to zero-emission vehicles (ZEVs) can, however partially, address the continuing disparate impact of U.S. waste-hauling practices.

Commercial Waste Zones and Exclusive Franchise Agreements

Some large cities have rethought the sprawling, inefficient networks of commercial waste collectors that service their territories, designing franchising programs that completely remake the commercial waste-hauling market. (Municipalities of all sizes have long had exclusive franchise arrangements for residential trash collection, but many cities require businesses to find their own haulers for their waste.) Los Angeles implemented an exclusive commercial waste franchise program across eleven zones of the city,[55] and New York City enacted a "commercial waste zones" law.[56] These franchising programs can have two broad benefits for reducing GHG emissions. First, they reduce miles traveled by waste-collection trucks by limiting the number of commercial waste haulers serving a neighborhood or zone to one or two, rather than allowing an unlimited number of trucks to crisscross a city to pick up waste.[57] Reducing truck miles also has important ancillary benefits like reducing air pollution and improving street safety. Second, they leverage contractual law—long contract terms in particular—to require or incentivize commercial waste franchisees to invest in recycling and composting infrastructure with more confidence that they will earn a financial return.

As with many reallocations of business opportunities and costs to customers, waste-carting franchise arrangements can attract litigation. In Los Angeles, commercial and multitenant building owners sued the city for the increased costs that the franchise system brought with it. One prominent lawsuit, brought by two building owners in coordination with the Apartment Owners Association of California, alleged that the franchise system was uncompetitive and allowed franchisees to tack on extra fees that amounted to a local tax that required voter approval pursuant to California Proposition 218.[58] The city agreed to a settlement that included a refund or waiver of more than $9 million in fees to building and business owners and to take over approximately $7 million in costs per year going forward.

Litigation also arose in New York City relating to a law ancillary to the city's commercial waste zones law. There, the city had enacted Local Law 152, or the "waste equity law," which decreased permissible waste capacity at several transfer stations located in environmental justice neighborhoods.[59] In addition to other claims, the petitioners, a group of waste and recycling facility operators, argued that the city had "impermissibly segmented" its environmental review of the waste equity law from review of the commercial waste zone overhaul for purposes of the city's environmental review law. The court rejected this argument, writing that "Local Law 152 was drafted in response to specific health and hazard concerns of overburdened communities housing more than their fair share of New York City's garbage . . . and thus, not connected to the Commercial Waste Zone Plan."[60] The petitioners also argued that Local Law 152 overstepped the city's state-delegated authority, was inconsistent with the city's state-approved solid waste management plan, and violated petitioners' substantive due process rights. The court rejected all claims.

Several other cases have parsed the terms of exclusive waste-carting franchises. For example, two cases in Reno, Nevada, alleged that the city's franchise agreement with Waste Management constituted price fixing or other noncompetitive behavior. These claims were held meritless, including by the federal District Court for the District of Nevada and the U.S. Court of Appeals for the Ninth Circuit, which noted that Nevada's franchise enabling law "clearly articulated and affirmatively expressed [as] Nevada policy" a city's right to grant a monopoly waste-hauling franchise.[61] Litigation in Oakland, California, examined franchise fees charged by the city to the waste franchisee, with the court holding that such fees must be "reasonably related to the value received from the government" and that amounts beyond that reasonable relationship could be considered taxes that the city does not have the authority to impose.[62]

In other instances, small waste haulers have sued their municipalities when new franchise agreements displaced their preexisting franchises or business; results have been mixed, with some haulers receiving damages.[63] The cases here vary with applicable state law, and several states have implemented legislative protections for displaced haulers.[64] Municipalities around the country, including in Reno and Oakland, continue to have franchise systems for residential and commercial waste hauling, and, unless disallowed by state law, this is generally an acceptable exercise of municipal authority.

Requiring Trash Haulers to Use Electric Trucks or Other Zero-Emission Vehicles

One seemingly straightforward—but legally fraught—way to reduce waste transport emissions is to transition trash haulers to all-electric (or otherwise zero-emission) fleets. Doing so can give rise to federal preemption scrutiny under the Clean Air Act (CAA) and the Energy Policy and Conservation Act (EPCA) (see chapter 4). A local law or other requirement that private waste haulers purchase or deploy electric trucks or other ZEVs would almost certainly be preempted by the CAA, EPCA, or both. An incentive for private haulers to use ZEVs would likely be legally permissible, as long as the incentive was not "so coercive as to" amount to a de facto mandate.[65] A local government may take advantage of a market participant exception to both the CAA and EPCA, allowing it to spend its own money or use its own property in the way it chooses. In other words, transitioning a *city-owned and -operated* waste-hauling fleet to ZEVs is far simpler—legally speaking—than requiring a *private* waste hauler to do so (see chapter 4).

THE GLOBAL CHAIN OF PRODUCTION AND CONSUMPTION

While the most direct policy impacts a city can have on its waste-related GHG emissions are achieved by reducing landfilled waste and limiting the emissions associated with other waste-processing and recycling operations, cities can play a role in requiring and incentivizing measures to reduce global GHG emissions resulting from the production of goods. Reducing the global production of items that end up as waste (e.g., packaging, disposable items) or finding ways to repurpose materials that are tossed after short- or long-term use reduces, at a global level, the waste that ends up landfilled as municipal solid waste.

A Circular Economy

A circular economy, in its ideal form, is one in which no material goes to waste; all materials are reused in some way and are often designed for such reuse or repurposing.[66]

U.S. cities are global economic actors; no city will be able to implement a completely circular economy on its own, and the benefits of moving toward circularity may accrue to the GHG emission inventories of other places. Still, there are legal tools that municipalities can use on their own to advance a circular economy both locally and globally. Some of the legal tools discussed in this chapter, like organic waste diversion requirements and bans on single-use plastic items,[67] encourage a more circular economy that "designs out"[68] sources of waste and finds secondary uses for waste that remains. Zero-waste goals are also tools aligned with creating a circular economy, as are requirements to limit construction waste or assess building life-cycle costs.

A municipality could also adhere to the principles of a circular economy in its operations and as a market participant; that is, by requiring waste diversion in its buildings and operations and by choosing contractors that divert waste.

Consumption-Based Emissions Accounting

Consumption-based emissions accounting offers an alternative to the sector-based accounting prescribed by the *Global Protocol for Community-Scale Greenhouse Gas Emission Inventories*. Under consumption-based accounting, a city counts in its emissions inventory "direct and life cycle GHG emissions of goods and services (including those from raw materials, manufacture, distribution, retail and disposal) and allocates GHG emissions to the final consumers of those goods and services."[69] This stands in contrast to the *Global Protocol* methodology of calculating the transportation, waste, and stationary energy emissions generated within a city's boundaries regardless of where a good or service is consumed. Modeling by C40 Cities has suggested that, if consumption-based emissions accounting methods were used, some U.S. cities could report more than three times as many GHG emissions as they currently do under the *Global Protocol*.[70] A small handful of U.S. cities and counties have performed consumption-based emissions analyses.[71]

There is little from a legal perspective that weighs in favor or against this approach. Consumption-based emissions accounting can help cities delineate the full life-cycle GHG impact of their residents and encourage the local production of goods (because production emissions don't "count" in a city's inventory if the goods are shipped away).[72] By exploring consumption-based emissions in parallel with sector-based emissions, a city could better develop the law and policy it needs to reduce its share of global emissions. A consumption-based emissions analysis could also help a city develop more equitable policy related to GHG emissions and waste reduction, because the econometric analysis performed to inventory consumption-based GHG emissions can be done at the neighborhood scale, allowing visibility into where in a city emissions are highest.[73] While cities will likely continue to use the *Global Protocol* approach, which aligns with their existing goals to reduce GHG emissions and their climate action planning, consumption-based accounting could serve as a helpful complement in demonstrating climate leadership and achieving deeper carbon reductions.

CONCLUSION

The years since the start of the COVID-19 pandemic have served to amplify the economic, public health, racial justice, and other needs of cities and their residents. The climate crisis poses a profoundly real set of challenges, given the physical, tangible, pervasive ways in which cities and day-to-day life must be remade to achieve local climate goals. However you see it, the imperative to reduce GHG emissions at the local level is intertwined with all aspects of city life.

The next several years are critical. The core challenge explored throughout this primer—how cities can resolve legal questions presented by their need to increase energy efficiency, electrify systems, and bring carbon-free energy into the electric grid—is both immediate (to meet the short- and mid-term targets necessary to preserve a chance of limiting global warming while society transitions to a carbon-free future) and long-lasting (the transition will not be enacted or completed any time soon). Local leaders with ambitious climate targets and commitments must untangle the layered legal frameworks that govern federal, state, and local authority; environmental pollution; energy policy; and much more. Most of our laws were not written to address the climate emergency. Some existing laws will shape the boundaries of city climate policy, whereas others will preempt certain policies altogether. The politics of climate and energy in the United States means that future preemption by state legislatures adds to the potential challenges.

Cities' leadership is vital to climate action. As we have seen in recent years, federal leadership is subject to political ebbs and flows, constrained by what can and cannot be done in Congress and by a Supreme Court majority skeptical of regulation, and has yet to fully rise to the task. Local governments are on the front lines of climate change and see the needs of communities and individuals on the ground not only to lessen the impacts of climate change but also to address pollution and its negative health impacts, create green jobs, remedy historical environmental racism, and create more accessible places to live and work.

Though some states have passed comprehensive climate legislation, cities are most often left to apply principles of existing bodies of law to the innovative policies in emerging domains. Building codes have become central to efforts to decarbonize buildings. Cities, constrained by federal and state authority over vehicles and highways, use land use and transportation-related authority to help drive a decrease in vehicle miles traveled and an increase in the adoption of electric vehicles. Cities are finding new ways to engage with the state public service commissions that set the terms for utility expansions and energy rate-making, two central elements of decarbonizing the energy system in an equitable way. And the long history of civil rights law and equal protection in the United States has implications in this new context as well.

This primer takes these various legal principles and transposes them into the local sphere of climate policy-making. In so doing, two broad themes become clear. First, local governments face significant legal obstacles to achieving their goals of reducing GHG emissions at both the state and federal levels. Second, local governments also have significant legal opportunities to address the climate challenge given their unique roles and sources of authority. We hope this primer will serve as a starting point for students, practitioners, policy-makers, and public officials seeking to understand the legal frameworks applicable to their climate policies and move forward on implementing policies that will both meet their communities' needs and survive legal challenges. For the future of cities, nothing could be more important.

NOTES

A NOTE ON TERMINOLOGY AND GLOSSARY

1. U.S. CONST., art. I, § 8, cl. 3.
2. "What Is Energy Justice?," Initiative for Energy Justice, accessed March 22, 2022, https://iejusa.org.
3. See, e.g., "Environmental Justice," U.S. Environmental Protection Agency, accessed March 22, 2022, https://www.epa.gov/environmentaljustice.
4. 23 U.S.C. § 101(a)(6).
5. 49 U.S.C. § 32901(a)(11).
6. *Agins v. City of Tiburon*, 447 U.S. 255, 260 (1980).
7. "Glossary of Terms," American Cities Climate Challenge, accessed March 22, 2022, https://cityrenewables.org/glossary/.

INTRODUCTION

1. "Check Out Where We Are Ready for 100%," Sierra Club, accessed April 28, 2023, https://www.sierraclub.org/ready-for-100/map?show=committed.
2. Samuel A. Markolf, Inês M. L. Azevedo, Mark Muro, and David G. Victor, *Pledges and Progress: Steps Toward Greenhouse Gas Emissions Reductions in the 100 Largest Cities Across the United States* (Washington, DC: Brookings Institution, October 2020), https://www.brookings.edu/wp-content/uploads/2020/10/FP_20201022_ghg_pledges_v4.pdf.
3. Wee Kean Fong, Mary Sotos, Michael Doust, Seth Schultz, Ana Marques, and Chang Deng-Beck, *Global Protocol for Community-Scale*

Greenhouse Gas Inventories: An Accounting and Reporting Standard for Cities, version 1.1 (Washington, DC: World Resources Institute, 2021), https://ghgprotocol.org/sites/default/files/standards/GPC_Full_MASTER_RW_v7.pdf.

4. C40 Cities Climate Leadership Group and C40 Knowledge Hub, "Why Cities Need to Advance Towards Zero Waste," C40 Cities, May 2019, https://www.c40knowledgehub.org/s/article/Why-cities-need-to-advance-towards-zero-waste?language=en_US.
5. U.S. CONST. art. I, § 8, cl. 3.
6. Jonathan D. Rosenbloom, *Outsourced Emissions: Why Local Governments Should Track and Measure Consumption-Based Greenhouse Gases*, 92 U. COLO. L. REV. 451 (2021).

1. CROSS-CUTTING LEGAL CONCEPTS

1. See, e.g., *City of New York v. State*, 86 N.Y.2d 286, 289–90 (N.Y. Ct. of Apps. 1995); *Pennington County v. State ex rel. Unified Judicial System*, 641 N.W.2d 127, 132 (S.D. Supr. Ct. 2002); *Town of Hooksett v. Baines*, 148 N.H. 625, 628 (N.H. Supr. Ct. 2002); *Palmer v. Inhabitants of Town of Sumner*, 133 Me. 337 (Me. Supr. Jud. Ct. 1935).
2. Indiana Publ. L. No. 180 (2021).
3. Laurie Reynolds, *Taxes, Fees, Assessments, Dues, and the "Get What You Pay For" Model of Local Government*, 56 FL. L. REV. 373, 412 (2004). Delineations are sometimes made between regulatory fees, such as licensing fees, and user fees, which pay the government for services it provides (e.g., trash pick-up). In some jurisdictions, a municipality may assess an impact fee or an exaction to require new development to "pay its own way." Different courts will apply different versions of the criteria outlined by Reynolds to different kinds of fees. See also *Emerson College v. City of Boston*, 391 Mass. 415, 424 (1984).
4. 5 McQuillin Mun. Corp. §§ 17.15 & 17.6 (3d ed.) (2022).
5. N.Y. Veh. & Traf. Law §1630 (2019).
6. Or. Rev. Stat. § 383.004(2) (2007) and OR. CONST. art. IX, § 3a.
7. Wash. Rev. Code § 36.73.020 (2010).
8. N.U. Mun. Home Rule L. § 23(2)(f).
9. See, e.g., *City of San Antonio, Tex. v. San Antonio Firefighters' Ass'n, Local 624*, 533 S.W.3d 527, 543 (Tex. App. 2017); *Russell City Energy Co., LLC v. City of Hayward*, 14 Cal. App. 5th 54, 222 Cal. Rptr. 3d 162 (2017);

The Lamar Co., LLC v. City of Columbia, 512 S.W.3d 774, 784 (Mo. Ct. App. 2016) and more.

10. U.S. CONST. art. VI, cl. 2.
11. Clean Air Act, 42 U.S.C. § 7401 et seq. (1970); Energy Policy and Conservation Act, 42 U.S.C. § 8251 et seq. (1975).
12. U.S. CONST. art. I, § 8, cl. 3 and *Wyoming v. Oklahoma*, 502 U.S. 437, 454 (1992).
13. *City of Philadelphia v. New Jersey*, 437 U.S. 617 (1978) (case was in the context of accepting out-of-state waste for landfilling and did not relate to GHG emissions).
14. U.S. CONST. amend. XIV, § 1.
15. *Cohen v. Rhode Island Turnpike and Bridge Authority*, 775 F.Supp.2d 439 (D.R.I. 2011); *Taxifleet Mgmt. LLC v. State of N.Y.*, Index No. 161920 /2018, Decision/Judgment, at 10 (June 25, 2019).
16. U.S. CONST. amend. IV.
17. See generally Amy E. Turner, *Legal Tools for Achieving Low Traffic Zones*, 50 Env. L. REP. 10329, 10338–40 (Apr. 2020).
18. See, e.g., *Naperville Smart Meter Awareness v. City of Naperville*, 900 F.3d 521 (7th Cir. 2018) (holding that the collection of smart meter data constitutes a search but that the search is reasonable given "the significant government interests in the program").
19. See generally Jonathan Rosenbloom, *Remarkable Cities and the Fight Against Climate Change: 43 Recommendations to Reduce Greenhouse Gases and the Communities That Adopted Them* (Washington, DC: Environmental Law Institute, 2020).
20. *Village of Euclid, Ohio v. Ambler Realty Co.*, 272 U.S. 365 (1926).
21. U.S. CONST. amend. V: " . . . nor shall private property be taken for public use, without just compensation."
22. *Kelo v. City of New London, Conn.*, 545 U.S. 469 (2005).
23. *Lucas v. South Carolina Coastal Council*, 505 U.S. 1003, 1019 (1992).
24. 11 McQuillin Mun. Corp. § 32:27 (3d ed.).
25. U.S. CONST. amend. V.
26. Chris Brown, "How High-Performance Leases Help Landlords and Tenants Achieve Sustainability Goals," Institute for Market Transformation, March 19, 2019, https://www.imt.org/how-high-performance-leases-give-landlords-and-tenants-something-to-agree-on/.
27. *Dockless Mobility Regulation*, Practical Law Government Practice Law Practice Notes w-017-6569 (2019), available on Westlaw.

28. Zhaoyu Kou, Xi Wang, Shun Fung (Anthony) Chiu, and Hua Cai, "Quantifying Greenhouse Gas Emissions Reduction from Bike Share Systems: A Model Considering Real-World Trips and Transportation Mode Choice Patterns," *Resources, Conservation and Recycling* 153 (2020): 104534, https://doi.org/10.1016/j.resconrec.2019.104534.

2. EQUITY

1. "Climate Resilience & Equity," Office of Climate Change, Sustainability and Resiliency, accessed April 15, 2022, https://resilientoahu.org/equity.
2. Chief Equity Officer Brion Oaks to Department Directors, memorandum, "Equity, Diversity, and Inclusion Trainings," October 22, 2019, City of Austin, "Memorandum," http://www.austintexas.gov/edims/pio/document.cfm?id=329993.
3. *City of Richmond v. J. A. Croson*, 488 U.S. 469, 509 (1989).
4. *Vill. of Arlington Heights v. Metro. Housing Dev. Corp.*, 429 U.S. 252, 259 (1977).
5. *Washington v. Davis*, 426 U.S. 229, 236 (1976).
6. *R.I.S.E., Inc. v. Kay*, 768 F. Supp. 1144, 1149-51 (E.D. Va. 1991), aff'd, 977 F.2d 573 (4th Cir. 1992).
7. See *Bean v. Southwestern Waste Mgmt. Corp.*, 482 F. Supp. 673, 680 (S.D. Tex. 1979); *R.I.S.E. v. Kay*, 977 F.2d 573, 1992 WL 295129, at *2, *4 (4th Cir. Oct. 15, 1992);
8. 42 U.S.C. § 2000d (1964).
9. *Alexander v. Sandoval*, 532 U.S. 275, 293 (2001).
10. *South Camden Citizens in Action v. New Jersey Dept. of Envtl. Prot.*, 274 F.3d 771, 790 (3d Cir. 2001); 42 U.S.C. § 1983 (1996).
11. *Vecinos para el Bienestar de la Comunidad Costera v. FERC*, 6 F.4th 1321 (D.C. Cir. 2021) and *Milwaukee Inner-City Congregations Allied for Home (MICAH) v. Gottlieb*, 944 F. Supp. 2d656 (W.D. Wis. 2013). See also Deborah N. Archer, *"White Men's Roads Through Black Men's Homes": Advancing Racial Equity Through Highway Reconstruction*, 73 VANDERBILT L. REV. 1259 (2020).
12. E.g., Cal. Env'tl Policy Act (CEQA), Cal. Pub. Res. Code § 21000 et seq. (1970).
13. Exec. Order No. 12898, 59 Fed. Reg. 7629 (Feb. 11, 1994).

14. Exec. Order. No. 14008, 86 C.F.R. 7620 (Jan. 27, 2021).
15. *Id.* § 223. The definition of *disadvantaged community*, and whether and how such definition should account for race, have attracted scrutiny.
16. Memorandum on Redressing Our Nation's and the Federal Government's History of Discriminatory Housing Practices and Policies, 86 C.F.R. 7487 (Jan. 26, 2021).
17. Aman Azhar, "Expansion of I-45 in Downtown Houston Is on Hold, for Now, in a Traffic-Choked, Divided Region," *Inside Climate News*, April 30, 2021, https://insideclimatenews.org/news/30042021/expansion-of-i-45-in-downtown-houston-is-on-hold-for-now-in-a-traffic-choked-divided-region/.
18. Wisconsin Department of Transportation, "WisDOT to Expand Review of I-94 East/West Project," news release, April 15, 2021, https://wisconsindot.gov/Documents/projects/by-region/se/94ew-study/2021/supplemental-release.pdf.
19. Michael S. Regan, administrator of U.S. Environmental Protection Agency, to Lori E. Lightfoot, mayor of Chicago, May 7, 2021.
20. "General Iron Move Halted as Biden EPA Weighs In," *Crain's Chicago Business*, May 7, 2021.
21. City of Chicago Department of Public Health to Hal Tolin, "Recycling Facility Permit Application," February 18, 2022, https://www.chicago.gov/content/dam/city/sites/rgm-expansion/documents/Final%20RMG%20permit%20denial%20letter%202.18.22%20with%20attachments.pdf.
22. Pub. L. No. 117-169 (Aug. 16, 2022).
23. Pub. L. No. 117-58 (Nov. 15, 2021).
24. N.J. Rev. Stat. § 13:92 (2020).
25. N.Y. Envt'l Conservation L. § 75-0109 (2020).
26. N.Y. City Admin. Code § 24-803 (2019).
27. Local Law 97 Carbon Trading Study Group, *Carbon Trading for New York City's Building Sector: Report of the Local Law 97 Carbon Trading Study Group to the New York City Mayor's Office of Climate & Sustainability* (New York: Guarini Center on Environmental, Energy & Land Use Law, 2021), https://policyintegrity.org/files/publications/2021-11-15_Guarini_-_Carbon_Trading_For_New_York_Citys_Building_Sector.pdf.
28. Va. Code Ann. § 10.1-1307(E)(3) (2021).

29. *Friends of Buckingham v. State Air Pollution Control Bd.*, 947 F.3d 68, 71 (4th Cir. 2020).
30. Cal. Env'tl Policy Act (CEQA), Cal. Pub. Res. Code § 21000 et seq. (1970).
31. N.Y. Envt'l Conserv. L. § 8-0101 et seq. (1976).
32. E.g., N.Y. City Local L. 78 (2021); City of Seattle, *Seattle 2035 Equity Analysis* (Seattle, WA: Office of Planning & Community Development, n.d.), https://www.seattle.gov/Documents/Departments/OPCD/OngoingInitiatives/SeattlesComprehensivePlan/2035EquityAnalysisSummary.pdf.
33. N.Y. Envt'l Conserv. L. §§ 75-0117, 75-0101(5) (2020).
34. 2021 Wash. Sess. L. S.5126 (2021).
35. Exec. Order. No. 14008, 86 C.F.R. 7620 (Jan. 27, 2021).
36. See David Madland and Terry Meginniss, "5 Ways State and Local Governments Can Make Climate Jobs Good Jobs," Center for American Progress, October 9, 2020, https://www.americanprogress.org/issues/economy/reports/2020/10/09/491226/5-ways-state-local-governments-can-make-climate-jobs-good-jobs/. For a more comprehensive look at developing "high-road" workforce policies, see Inclusive Economics, *High-Road Workforce Guide for City Climate Action* (Port Washington, WI: Urban Sustainability Directors Network, 2021), 16.
37. The city of Milwaukee, for example, has highlighted the need for funding for climate and economic equality efforts and the "severe revenue constraints imposed [on the city] by the state." Task Force Working Groups and Legislative Reference Bureau, *City-County Task Force on Climate and Economic Equity: Preliminary Report* (Milwaukee, WI: City of Milwaukee, March 2020), 37.
38. N.Y. CONST. art. VII, § 10a.
39. OR. CONST. art. IX, § 3a.
40. New York City's 2006 Solid Waste Management Plan, for example, was aimed at reducing the transportation of solid waste by trucking in favor of rail and marine transfer and siting transfer stations in an equitable way. City of New York and Department of Sanitation, *Comprehensive Solid Waste Management Plan* (New York: City of New York, September 2006), https://www1.nyc.gov/assets/dsny/site/resources/reports/solid-waste-management-plan.
41. Somerville, Ma., Ordinance 96 (Dec. 12, 2019).

3. BUILDINGS

1. See, e.g., "Greenhouse Gas Emissions Interactive Dashboard," C40 Cities, accessed March 22, 2022, https://www.c40knowledgehub.org/s/article/C40-cities-greenhouse-gas-emissions-interactive-dashboard?language=en_US.
2. See, e.g., James H. Williams, Benjamin Haley, Frederick Kahrl, Jack Moore, Andrew D. Jones, Margaret S. Torn, and Haewon McJeon, *Pathways to Deep Decarbonization in the United States*, U.S. report of the Deep Decarbonization Pathways Project of the Sustainable Development Solutions Network and the Institute for Sustainable Development and International Relations, revision with technical supplement (San Francisco: Energy and Environmental Economics, Inc., 2015), http://usddpp.org/downloads/2014-technical-report.pdf.
3. See, e.g., James Mandel and Laurie Stone, "Making Our Existing Buildings Zero Carbon: A Three-Pronged Approach," Rocky Mountain Institute, December 4, 2019, https://rmi.org/making-our-existing-buildings-zero-carbon-a-three-pronged-approach/.
4. "Scope 1 and Scope 2 Inventory Guidance," U.S. Environmental Protection Agency, accessed March 15, 2022, https://www.epa.gov/climateleadership/scope-1-and-scope-2-inventory-guidance.
5. 42 U.S.C. § 6297(b) (2010).
6. 42 U.S.C. §§ 6832(15), 6833(a); Lee Paddock and Caitlin McCoy, "New Buildings," in *Legal Pathways to Deep Decarbonization in the United States*, ed. Michael B. Gerrard and John C. Dernbach (Washington, DC: Environmental Law Institute, 2019).
7. Maura Healey, attorney general of Massachusetts, to Patrick J. Ward, town clerk, and Linda Goldburgh, assistant town clerk, Town of Brookline, July 21, 2020, re Brookline Special Town Meeting of November 19, 2019—Case # 9725 Warrant Article # 21 (General), https://www.brooklinema.gov/DocumentCenter/View/22350/Brookline-9752S_DIS_final.
8. *Azzar v. City of Grand Rapids*, 2005 WL 2327076 (2005).
9. 35 Pa. Stat. § 7210.503(i).
10. N.Y. Exec. L. § 379 (2017).
11. Mass. S.B. 9 (2021).

12. Ca. Energy Comm'n 2022 Building Energy Efficiency Standards (approved by Cal. Bldg. Stds. Comm'n in Dec. 2021).
13. 42 U.S.C. §§ 6201 et seq.
14. 40 C.F.R. pt. 60 (2023).
15. 42 U.S.C. § 6297(b).
16. *Id.*
17. 42 U.S.C. § 6297(d). For more about exceptions to EPCA preemption, see Peter Ross, *Appliance & Equipment Efficiency Standards: A Roadmap for State & Local Action* (Sabin Center for Climate Change L., Columbia L. Sch. Working Paper, July 2017).
18. *California Restaurant Ass'n v. City of Berkeley*, No. 21-16278 (9th Cir. 2023).
19. 42 U.S.C. § 6297(e).
20. Ross at 22–23.
21. *Id.* at 12–20.
22. 42 U.S.C. § 6297(f).
23. 42 U.S.C. § 6297(f)(3).
24. *Air Conditioning, Heating and Refrigeration Institute v. City of Albuquerque*, 2008 WL 5586316 (U.S. Dist. Ct. D. N.M. 2008).
25. *Id.* at *12.
26. *Building Industry Ass'n of Wash. v. Wash. State Building Code Council*, 683 F.3d 1144 (9th Cir. 2011).
27. *Id.*
28. 780 C.M.R. ch. 115AA (2018).
29. Vermont Res. Bldg. Stds., 5th ed. (2020).
30. "NYStretch Energy Code: 2020 Outreach, Training and Resources," New York State Energy Research and Development Authority, accessed March 22, 2022, https://www.nyserda.ny.gov/All-Programs/Programs/Energy-Code-Training/NYStretch-Energy-Code-2020.
31. "Green Communities Division," Mass.gov, accessed March 22, 2022, https://www.mass.gov/orgs/green-communities-division.
32. Jessica Gable, "California's Cities Lead the Way on Pollution-Free Homes and Buildings," Sierra Club, last updated February 14, 2023), https://www.sierraclub.org/articles/2021/07/californias-cities-lead-way-pollution-free-homes-and-buildings. Seattle also requires all-electric construction through its building code.

33. On the variety of local code provisions in California, see Caitlin McCoy, *The Legal Dynamics of Local Limits on Natural Gas Use in Buildings* (Cambridge, MA: Harvard Law School Environmental & Energy Law Program, June 8, 2020), http://eelp.law.harvard.edu/wp-content/uploads/The-Legal-Dynamics-of-Local-Limits-on-Natural-Gas-Use-in-Buildings.pdf.
34. Anthony Derrick, "Mayor Durkan Announces Ban on Fossil Fuels for Heating in New Construction to Further Electrify Buildings Using Clean Energy," Office of the Mayor, Seattle, December 3, 2020, https://durkan.seattle.gov/2020/12/mayor-durkan-announces-ban-on-fossil-fuels-for-heating-in-new-construction-to-further-electrify-buildings-using-clean-energy/.
35. N.Y. City Local L. 154 (2021).
36. Amy Turner, "Municipal Natural Gas Bans: Round 1," *Climate Law* (Sabin Center blog), January 9, 2020, https://blogs.law.columbia.edu/climatechange/2020/01/09/municipal-natural-gas-bans-round-1/.
37. *California Restaurant Ass'n v. City of Berkeley*, No. 21-16278 (9th Cir. 2023).
38. Healey to Ward and Goldburgh.
39. Brookline, Mass. Warrant Articles 25 and 26 (2021).
40. See, e.g., Ariz. H.B. 2686 (2020); Louisiana S.B. 492 (2020); Okla. H.B. 3619 (2020). For a full list, see Alejandra Mejia Cunningham and Kimi Narita, "Gas Interests Threaten Local Authority," Natural Resources Defense Council, last updated June 12, 2023, https://www.nrdc.org/experts/alejandra-mejia/gas-interests-threaten-local-authority-6-states.
41. N.Y. State. Pub. Serv. L. § 30.
42. For more, see Emily Pontecorvo, "Does Your State Want to Cut Carbon Emissions? These Old Laws Could Be Standing in the Way," *Grist*, August 10, 2020, https://grist.org/energy/does-your-state-want-to-cut-carbon-emissions-these-old-laws-could-be-standing-in-the-way-buildings-heat-pumps/.
43. *Gallaher v. Town of Windsor*, Docket No. SCV-265553 (Cal. Super. Ct. Nov. 19, 2019) and *Windsor Jensen Land Company, LLC v. Town of Windsor*, Docket No. SCV-265583 (Cal. Super. Ct. Nov. 22, 2019).
44. *Gallaher v. City of Santa Rosa*, Docket No. SCV-265711 (Cal. Super. Ct. Dec. 17, 2019).

45. *Gallaher v. City of Santa Rosa*, Ruling on Petition for Writ of Mandate and Complaint for Declaratory and Injunctive Relief (Apr. 22, 2021) and Notice of Appeal (June 21, 2021).
46. See, e.g., Sarah Ravani, "Berkeley Becomes the First U.S. City to Ban Natural Gas in New Homes," *San Francisco Chronicle*, July 21, 2019, https://www.sfchronicle.com/bayarea/article/Berkeley-becomes-first-U-S-city-to-ban-natural-14102242.php.
47. *Cal. Restaurant Ass'n v. City of Berkeley*, 4:19-cv-07668 (U.S. Dist. Ct. N.D. Cal. 2019).
48. *Cal. Restaurant Ass'n v. City of Berkeley*, Order Granting in Part and Denying in Part Motion to Dismiss at 15 (July 6, 2021).
49. Town of Watertown Zoning Ordinance, art. VIII § 8.05.
50. City of Cambridge, Mass. zoning Ordinance, art. 22.000.
51. City of Miami, Fla. Zoning Code § 3.13.
52. City of Minneapolis, Minn. Code of Ordinances § 549.220(12) (2016). See Brandon Hanson, "Energy and Water Efficiency," in *Sustainable Development Code: Climate Change*, ed. Jonathan Rosenbloom and Christopher Duerkson (Washington, DC: Environmental Law Institute, 2020), 57–58.
53. City of Miami Beach, Fl. Code of Ordinances § 1333 (2016). See Kerrigan Owens, "Third-Party Certification Requirements," in *Sustainable Development Code: Climate Change*, ed. Jonathan Rosenbloom and Christopher Duerkson (Washington, DC: Environmental Law Institute, 2020), 178.
54. Spur, Tex. Ordinance 667. See Kerrigan Owens, "Tiny Homes and Compact Living Spaces," in *Sustainable Development Code: Climate Change*, ed. Jonathan Rosenbloom and Christopher Duerkson (Washington, DC: Environmental Law Institute, 2020), 47.
55. Ann Arbor, Mich. Code of Ordinances § 5:10.2 4(d) (2016). See Tyler Adams, "Accessory Dwelling Units," in *Sustainable Development Code: Climate Change*, ed. Jonathan Rosenbloom and Christopher Duerkson (Washington, DC: Environmental Law Institute, 2020), 3–4.
56. Somerville, Mass. Zoning Ordinance ch. 3.
57. Andrea McCardle, *Local Green Initiatives: What Local Governance Can Contribute to Environmental Defenses Against the Onslaughts of Climate Change*, 28 FORDHAM ENVT'L L. REV. 102, 108 (2016). See *Whitman v. Am. Trucking Ass'ns*, 531 U.S. 457 (2001).

58. See "LEED Credit Library," U.S. Green Building Council, accessed March 22, 2022, https://www.usgbc.org/credits?Version=%22v4.1%22&Rating+System=%22New+Construction%22.
59. Energy Star (website), https://www.energystar.gov (accessed April 27, 2023).
60. See, e.g., Boston Zoning Code art. 37 (2007) and Washington, D.C. Green Building Act (2012).
61. Boston Zoning Code, art. 37 § 37–4 (2007).
62. See, e.g., Houston, Tex. Green Building Res. (2004).
63. Washington, D.C. Green Buildings Act (2006).
64. Nancy E. Shurtz, *Eco-friendly Building from the Ground Up: Envt'l Initiatives and the Case of Portland, Ore*, 27 J. ENVT'L L. & LITIG. 237, 268 n.136 (2012).
65. *Chicago Sustainable Development Policy Handbook*, https://www.chicago.gov/city/en/depts/dcd/supp_info/sustainable_development/chicago-sustainable-development-policy-handbooko.html (accessed May 1, 2023). See also Carl J. Circo, *Using Mandates and Incentives to Promote Sustainable Construction and Green Building Projects in the Private Sector: A Call for More State Land Use Policy Initiatives*, 112 PENN. ST. L. REV. 738, 761 (2008).
66. "Building Energy Use Benchmarking," U.S. Department of Energy Office of State and Community Energy Programs, accessed March 22, 2022, https://www.energy.gov/eere/slsc/building-energy-use-benchmarking.
67. "Benchmarking Your Building Using Energy Star Portfolio Manager," Energy Star, https://www.energystar.gov/buildings/benchmark, accessed May 1, 2023.
68. "Energy Benchmarking and Transparency Benefits," Institute for Market Transformation, accessed March 22, 2022, https://www.imt.org/resources/fact-sheet-energy-benchmarking-and-transparency-benefits/.
69. N.Y. City Local L. 33 (2018) and Local L. 95 (2019), as codified by N.Y. City Admin. Code § 28–309.12. New York is the first city in the United States to use a letter grade as a form of public disclosure, but the system draws on similar requirements in Europe. Directive 2010/31/EU art. 11. See Danielle Spiegel-Feld, *Building Demand for Efficient Buildings: Insights from the EU's Energy Disclosure Regime* (New York: Guarini Center, April 2016), https://guarinicenter.org/wp-content/uploads/2016/05/Building-Demand-for-Efficient-Buildings_April-2016-1.pdf.
70. Chicago Municipal Code § 18-14-101.3 et seq (2019).

71. Tyler Adams, "Energy Benchmarking, Auditing and Upgrading," in *Sustainable Development Code: Climate Change*, ed. Jonathan Rosenbloom and Christopher Duerkson (Washington, DC: Environmental Law Institute, 2020), 126.
72. Minneapolis Code of Ordinances § 248.75 (2019).
73. Portland Ordinance No. 188143 (2018), City Code ch. 17.108.
74. See, e.g., Bureau of Planning and Sustainability, *Report to Portland City Council on Residential Energy Performance Rating and Disclosure (Ordinance No. 188143)* (Portland, OR: Bureau of Planning and Sustainability, October 2020), 3–4, https://www.portland.gov/sites/default/files/2020/report-to-portland-city-council-on-residential-energy-performance-rating-and-disclosure-ordinance-no.-188143.pdf.
75. *Naperville Smart Meter Awareness v. City of Naperville*, 900 F.3d 521, 528–29 (7th Cir. 2018). See also *Klein v. Met Ed*, 2020 WL 94077 (U.S. Dist. Ct. M.D. Penn. 2020).
76. Brandon J. Murrill, Edward C. Liu, and Richard M. Thompson II, *Smart Meter Data: Privacy and Cybersecurity*, Congressional Research Service Report (Washington, DC: Library of Congress Congressional Research Service, February 3, 2012).
77. 18 U.S.C. §§ 2510–2523 (1986).
78. 18 U.S.C. § 1030 (2020).
79. A.B. 375, 2017–2018 Leg. (Cal. 2018).
80. Steven Nadel and Adam Hinge, *Mandatory Building Performance Standards: A Key Policy for Achieving Climate Goals* (Washington, DC: American Council for an Energy-Efficient Economy, June 2020), 1, https://www.aceee.org/sites/default/files/pdfs/buildings_standards_6.22.2020_0.pdf.
81. For examples, see Clean Energy DC Omnibus Amendment Act of 2018, D.C. L. 22–257 (2019); New York Local L. 97 (2019); St. Louis Board Bill No. 219 (2020); Boston Building Emissions Reduction and Disclosure Ordinance (2021).
82. Reno Ordinance No. 6493 (2019); Reno Admin. Code §§ 14.30.011 et seq. (2019).
83. Boulder Ordinance Nos. 7724, 7725, and 7726 (2010), also known as "SmartRegs."
84. As of this writing, a case challenging the validity of New York City's building performance standard, Local Law 97, brought by a group of

building owners and cooperative board representatives is outstanding. New York City's government has filed a motion to dismiss. See *Glen Oaks Village Owners, Inc. v. City of New York*, Index No. 15437/2022, Complaint (May 28, 2022).

85. N.Y. City Local L. 97 (2019).
86. Mich. Comp. Ls. § 324.5542(1) (1995).
87. Cal. Health & Safety Code §§ 40000 et seq (1975).
88. Va. Code § 10.1–1321 (1994).
89. Minn. Stats. § 116.07(2) (2022).
90. Laurie Reynolds, *Taxes, Fees, Assessments, Dues, and the "Get What You Pay For" Model of Local Government*, 56 FL. L. REV. 373, 412 (2004).
91. Reynolds at 384.
92. 5 McQuillin Mun. Corp. § 17.15 (3d ed.)(2022).
93. See, e.g., N.Y. CONST. art. VIII.
94. See chapter 1 for further discussion of municipal fees and taxes.
95. *Id.*
96. N.Y. State Ls. (1974), ch. 574 § 4.
97. N.Y. City Local L. 97 (2019) § 6 (adding a new art. 321 to the N.Y. City Admin. Code).
98. Reno Ordinance No. 6493 (2019); Reno Admin. Code §§ 14.30.011 et seq. (2019); Nadel and Hinge, *Mandatory Building Performance Standards*, 13.
99. N.Y. City Local L. 97 (2019) and N.Y. City Local L. 116 (2020). New York State law was later amended to limit rent increases attributable to major capital improvements, further mitigating this concern.
100. See, e.g., N.Y. City Local L. 87 (2009). See also James Charles Smith, "Existing Buildings," in *Legal Pathways to Deep Decarbonization in the United States*, ed. Michael B. Gerrard and John C. Dernbach (Washington, DC: Environmental Law Institute, 2019), 286.
101. N.Y. City Local L. 94 (2019).
102. "NYC CoolRoofs," accessed March 22, 2022, https://www1.nyc.gov/nycbusiness/article/nyc-coolroofs.
103. For more on Boulder's SmartRegs requirements, see Alisa Petersen and Radhika Lalit, *Better Rentals, Better City: Smart Policies to Improve Your City's Rental Housing Energy Performance* (Basalt, CO: Rocky Mountain Institute, 2018), http://rmi.org/wp-content/uploads/2018/05/Better-Rentals-Better-City_Final3.pdf.

104. Austin, Tex. Code of Ordinances § 6-7-2 (2011).
105. San Francisco, Cal. Housing Code chs. 12 & 12A (2022).
106. See Smith, *Existing Buildings*, 287.
107. Kyle Massner, *Property Assessed Clean Energy Program*, in *Sustainable Development Code: Climate Change*, ed. Jonathan Rosenbloom and Christopher Duerkson (Washington, DC: Environmental Law Institute, 2020), 84.
108. Chris Brown, "How High-Performance Leases Help Landlords and Tenants Achieve Sustainability Goals," Institute for Market Transformation, May 19, 2019, https://www.imt.org/how-high-performance-leases-give-landlords-and-tenants-something-to-agree-on/.
109. Shurtz, *Eco-friendly Building*, 237, 333–34.
110. Andrew Feierman, *What's in a Green Lease? Measuring the Potential Impact of Green Leases in the U.S. Office Sector* (Washington, DC: Institute for Market Transformation, May 2015, https://www.imt.org/wp-content/uploads/2018/02/Green_Lease_Impact_Potential.pdf.
111. See, e.g., City of New York, *A Model Energy Aligned Lease Provision* (New York: City of New York, September 21, 2011), http://www.nyc.gov/html/planyc2030/downloads/pdf/energy_aligned_lease_official_packet.pdf.
112. "New Buildings: Embodied Carbon," Architecture 2030, accessed March 22, 2022, https://architecture2030.org/new-buildings-embodied/.
113. See Preservation Green Lab, *The Greenest Building: Quantifying the Environmental Value of Building Reuse* (Washington, DC: National Trust for Historic Preservation, 2011), vi, https://living-future.org/wp-content/uploads/2022/05/The_Greenest_Building.pdf, noting that "it can take between ten and eighty years for a new, energy-efficient building to overcome, through more efficient operations, the negative climate change impacts that were created during the construction process."
114. See, e.g., *St. George Greek Orthodox Cathedral v. Fire Dept. of Springfield*, 462 Mass. 120, 126 (2012).
115. Pasadena Municipal Code ch. 8.62 (2014).
116. Carl J. Circo, *Using Mandates and Incentives to Promote Sustainable Construction and Green Building Projects in the Private Sector: A Call for More State Land Use Policy Initiatives*, 112 PENN. ST. L. REV. 731, 752 (2008).
117. Jeremy Hays, Minna Toloui, Manisha Rattu, and Kathryn Wright, *Equity and Buildings: A Practical Framework for Local Government*

Decision Makers (Port Washington, WI: Urban Sustainability Directors Network, June 2021), https://www.usdn.org/uploads/cms/documents/usdn_equity_and_buildings_framework_-_june_2021.pdf.

4. REDUCING TRANSPORTATION-RELATED GREENHOUSE GAS EMISSIONS

1. "Sources of Greenhouse Gas Emissions," U.S. Environmental Protection Agency, https://www.epa.gov/ghgemissions/sources-greenhouse-gas-emissions accessed May 1, 2023. Data are for 2021.
2. See self-reported GHG data by sector and subsector of the North American member cities of C40 Cities, "a global network of mayors" focused on local climate policy and action, https://www.c40.org.
3. EPCA § 509(a), 49 U.S.C. § 32919(a).
4. *Metropolitan Taxicab Board of Trade v. City of New York*, 615 F.3d 152 (2d Cir. 2010), cert. denied, 562 U.S. 1264 (2011); *Ophir v. City of Boston*, 647 F. Supp. 2d 86 (D. Mass. 2009).
5. CAA § 209, 42 U.S.C. § 7543(a) (2010).
6. *Engine Manufacturers Ass'n v. South Coast Air Quality Management District*, 541 U.S. 246, 255 (2004).
7. *Id.*
8. *Id.*
9. *Green Alliance Taxi Cab Ass'n, Inc. v. King County*, No. C08-1048RAJ, 2010 WL 2643369 (W.D. Wash. June 29, 2010).
10. *Ass'n of Taxicab Operators USA v. City of Dallas*, 720 F.3d 534, 541 (5th Cir. 2013).
11. *Metropolitan Taxicab Board of Trade v. City of New York*, 615 F.3d 152 (2d Cir. 2010), cert. denied, 562 U.S. 1264 (2011).
12. U.S. CONST. art. I, § 8, cl. 3.
13. *Wyoming v. Oklahoma*, 502 U.S. 437, 454 (1992); *Or. Waste Sys., Inc. v. Department of Envt'l Quality of Or.*, 511 U.S. 93 (1994).
14. *Philadelphia v. New Jersey*, 437 U.S. 617, 624 (1978).
15. *Pike v. Bruce Church, Inc.*, 397 U.S. 137, 142 (1970).
16. *Metropolitan Taxicab I*, No. 08 Civ. 7837 (PAC), 2008 WL 4866021, at *7, 11–12 (S.D.N.Y. Oct. 31, 2008).
17. *Engine Mfrs. Ass'n v. South Coast Air Quality Mgmt. Dist.*, 498 F.3d 1031, 1040 (9th Cir. 2007) (*Engine Mfrs. II*).

18. *Tocher v. City of Santa Ana*, 219 F.3d 1040, 1049 (9th Cir. 2000).
19. *Engine Mfrs. II* at 1040.
20. 14 Cal. Code of Regs. § 15000 et seq (2023).; 6 N.Y. Codes, Rules & Regs. pt. 617 (2018).
21. See, e.g., *Council of Chelsea Block Ass'ns v. City of N.Y. Dep't of Transp.*, No. 156153/19 (N.Y. Sup. Ct. June 20, 2019); *14th St. Coalition v. City of N.Y. Dep't of Transp.*, No. 159030/18 (N.Y. Sup. Ct. Sept. 28, 2018).
22. *Ass'n of Taxicab Operators USA v City of Dallas*, 720 F.3d 534, 541 (5th Cir. 2013).
23. *Green Alliance Taxi Cab Ass'n, Inc. v. King County*, No. C08-1048RAJ, 2010 WL 2643369 (W.D. Wash. June 29, 2010).
24. Amy L. Stein and Joshua Fershée, "Light-Duty Vehicles," in *Legal Pathways to Deep Decarbonization in the United States*, ed. Michael B. Gerrard and John C. Dernbach (Washington, DC: Environmental Law Institute, 2019)., 357–66
25. Catherine Morehouse, "Should EV Charging Stations Be Regulated as Utilities? Kentucky Joins Majority in Saying No," *Utility Dive*, June 17, 2019, https://www.utilitydive.com/news/should-ev-charging-stations-be-regulated-as-utilities-kentucky-joins-major/556972/.
26. *Electronic Investigation of Comm'n Jurisd. Over Electric Vehicle Charging Stns.*, Case No. 2018–00372, Kent. Publ. Serv. Comm'n (2008).
27. Wash. Utilities and Transp. Comm'n, ESHB 1853, RCW 80.28.360 (2019); Colo. SB 19–077 § 2, amending CRS § 40-1-103.3(6) (2019). See Mark Detsky, Gabriella Stockmayer, *Electric Vehicles: Rolling over Barriers and Merging with Regulation*, 40 WM. & MARY ENVTL. L. & POL'Y REV. 477 (2016).
28. Trip Pollard, "Transforming Transportation Demand," in *Legal Pathways to Deep Decarbonization in the United States*, ed. Michael B. Gerrard and John C. Dernbach (Washington, DC: Environmental Law Institute, 2019), 309.
29. Pollard, "Transforming Transportation Demand," 309.
30. See "Drive Clean Seattle – FAQs," Seattle Office of Sustainability, https://www.seattle.gov/documents/departments/ose/dcs_faq_final.pdf (accessed May 1, 2023). See *Legal Pathways to Deep Decarbonization in the United States*, ed. Michael B. Gerrard and John C. Dernbach (Washington, DC: Environmental Law Institute, 2019), 368.

31. "Plug-In Austin Electric Vehicles," Austin Energy, accessed June 15, 2020, https://austinenergy.com/green-power/plug-in-austin.
32. See, e.g., Seattle Ordinance 125815 (2019); Fort Collins. Mun. Code Sec. 5–30 § E3401.5 (2019).
33. "Energy Code Stringency," American Council for an Energy-Efficient Economy, State and Local Policy Database, https://database.aceee.org/city/energy-code-stringency (accessed May 1, 2023).
34. In Massachusetts, for example, municipalities are expressly prohibited from including requirements regulated in the building energy code (which is set at the state level) in local zoning ordinances. M.G.L. ch. 40A § 3.
35. See, e.g., the stretch code of Massachusetts, 780 C.M.R. Appendix 115. AA (2018).
36. Salt Lake City Code ch. 21A.44.040.B (2019).
37. City of Chelan Mun. Code § 17.63 (2018).
38. Ordinance of the City Council of the City of Petaluma Amending the Text of the Implementing Zoning Ordinance, Ordinance 2300 N.C.S., to Modify Chapter 4 . . . (2021).
39. For example, port authorities in California have had to consider the state's public trust doctrine in determining where and for whom chargers can be installed.
40. U.S. CONST. amend. V.
41. Danielle Spiegel-Feld, *Local Law 97: Emissions Trading for Buildings?*, 94 N.Y.U. L. REV. 327, 341 (2019), citing Richard Briffault & Laurie Reynolds, STATE AND LOCAL GOVERNMENT LAW 581 (2016).
42. Spiegel-Feld at 342.
43. Infrastructure Investment and Jobs Act, Pub. L. 117–58 (2021).
44. Colo. H.B. HB19-1298 (2019).
45. Wash. Rev. Code § 81.72.210 (2020).
46. Cambridge, Mass. Ordinance. 4148 (2020).
47. Ordinance of the City Council of the City of Petaluma Amending the Text of the Implementing Zoning Ordinance, Ordinance 2300 N.C.S., to Modify Chapter 4 . . . (2021).
48. Faith E. Pinho, "Sixteen Gas Stations for 60,000 People? That's Enough, Petaluma Says," *Los Angeles Times*, March 4, 2021, https://www.latimes.com/california/story/2021-03-04/sixteen-gas-stations-for-60-000-people-thats-enough-petaluma-says.

49. Andrea Hudson Campbell, Avi B. Zevin, and Keturah A. Brown, "Heavy-Duty Vehicles and Freight," in *Legal Pathways to Deep Decarbonization in the United States*, ed. Michael B. Gerrard and John C. Dernbach (Washington, DC: Environmental Law Institute, 2019), 422.
50. Hudson Campbell, Zevin, and Brown, "Heavy-Duty Vehicles and Freight," 406.
51. See, e.g., Stein and Fershée, "Light-Duty Vehicles," 377–78.
52. Peter Plastrik and John Cleveland, *Game Changers: Bold Actions by Cities to Accelerate Progress Toward Carbon Neutrality*, ed. Michael Shank and Johanna Partin (Copenhagen: Carbon Neutral Cities Alliance, 2018).
53. 49 U.S.C. § 14501(c)(1) (2015).
54. *Pike v. Bruce Church, Inc.*, 397 U.S. 137, 142 (1970).
54. 42 U.S.C. §§ 4321–4370h (1970).
56. Or. Rev. Stat. § 383.004(2) (2007).
57. N.Y. Veh. & Traf. L. § 1630 (2019).
58. N.Y. Veh. & Traf. L. § 44-C (2019).
59. Wash. Rev. Code § 36.73.020 (2010).
60. *Christensen v. City of Pocatello*, 142 Idaho 132, 139 (Idaho 2005).
61. *Cohen v. City of Hartford*, 244 Conn. 206, 219 (Conn. 1998).
62. *Cohen v. City of Hartford* at 219.
63. See, e.g., Ark. Code §§ 12-12-1801 to -1808 (2013); Cal. Veh. Code § 2413 (2011) and Cal. Civ. Code §§ 1798.29 (2020), 1798.90.5 (2016); Colo. Rev. Stat. § 24-72-113 (2014); Fla. Stat. § 316.0777 (2019); Ga. Code § 35-1-22 (2018); Me. Rev. Stat. Ann. tit. 29-A, § 2117-A(2) (2019); Md. Code Ann., Pub. Safety § 3-509 (2019); Minn. Stat. §§ 13.82, 13.824, 626.8472 (2015); Mont. Code Ann. §§ 46-5-117 to -119 (2017); Neb. Rev. Stat. §§ 60-3201 to -3209 (2018); N.H. Rev. Stat. Ann. §§ 261.75-b (2016), 236.130 (2014); N.C. Gen. Stat. §§ 20-183.30 to .32 (2015); Tenn. Code § 55-10-302 (2014); Utah Code Ann. §§ 41-6a-2001 to -2005 (2018); Vt. Stat. Ann. tit. 23, §§ 1607, 1608 (2018); "Aggregated by Automated License Plate Readers: State Statutes," National Conference of State Legislatures, March 15, 2019, http://www.ncsl.org/research/telecommunications-and-information-technology/state-statutes-regulating-the-useof-automated-license-plate-readers-alpr-or-alpr-data.aspx.
64. See, e.g., *Neal v. Fairfax County Police Dept.*, 295 Va. 334, 346 (Va. 2018); *Gannett Co., Inc. v. County of Monroe*, 47 Misc. 3d 898, 905 (N.Y. Sup. Ct. 2015).

65. A.B. 375, 2017–2018 Leg. (Cal. 2018).
66. See, e.g., *Washington State Road Usage Charge Steering Committee, Steering Committee Report for the WA RUC Pilot Project* at 125–26, recommending updates to Washington's list of statutory exceptions to its public records disclosure law (Wash. Rev. Code §42.56.010(3)) (2022) to include mileage data.
67. See, e.g., *City & County of San Francisco v. Uber Techs., Inc.*, 36 Cal. App. 5th 66, 76 (Cal. Ct. App. 2019); *Lyft, Inc. v. City of Seattle*, 190 Wash. 2d 769 (Wash. 2018); *Rasier, LLC v. New Orleans*, 222 So. 3d 806, 813 (La. Ct. App. 2017); *City of Columbus v. Lyft, Inc.*, 22 N.E.3d 304 (Franklin County Mun. Ct. 2014); *Carniol v. N.Y. City Taxi & Limousine Comm'n*, 42 Misc. 3d 199, 209 (N.Y. Sup. Ct. 2013).
68. Alissa Walker, "Kansas City Becomes First Major U.S. City to Make Public Transit Free," *Curbed*, December 6, 2019, https://archive.curbed.com/2019/12/6/20998617/kansas-city-missouri-free-public-transportation.
69. See, e.g., "Transit-Oriented Development," Federal Transit Administration, last updated April 11, 2019, https://www.transit.dot.gov/TOD.
70. Chicago Zoning Code §§ 17-3-0403-B.
71. E.g., Bloomington Code of Ordinances § 19.29 (2018).
72. See, e.g., Rachel Weinberger, "Death by a Thousand Curb-Cuts: Evidence on the Effect of Minimum Parking Requirements on the Choice to Drive," *Transport Policy* 20 (March 2012): 93–102.
73. Rosenbloom at 157.
74. E.g., Portland., Ore. City Code § 33.366.115; Hartford, Conn. Zoning Regs. § 7.2.2(B)(2018).
75. Rosenbloom at 152.
76. Rosenbloom at 118.
77. Complete Streets Act of 2019, U.S. H.R. 3663 § 2(a)(3)(A) (2019) (bill referred to committee in July 2019).
78. See, e.g., "Vision Zero Communities," Vision Zero Network, https://visionzeronetwork.org/resources/vision-zero-cities/, accessed May 1, 2023.
79. Joseph Hollingsworth, Brenna Copeland, and Jeremiah X Johnson, "Are E-Scooters Polluters? The Environmental Impacts of Shared Dockless Electric Scooters," *Environmental Research Letters* 14, no. 8 (2019): 084031, https://doi.org/10.1088/1748-9326/ab2da8.
80. *Dockless Mobility Regulation*, Practical Law Government Practice Law Practice Notes w-017-6569 (2019), available on Westlaw.

81. 42 U.S.C. § 12101, et seq. (2008).
82. Greg Moran, "Disability Advocates Target City of San Diego and Scooter Companies in Lawsuit," *Los Angeles Times*, January 12, 2019, https://www.latimes.com/local/lanow/la-me-ln-ada-suit-scooters-san-diego-20190112-story.html; Andy Mannix, "Disability Rights Advocate Sues Rental Scooter Companies, Alleging Blocked Sidewalks," *Minneapolis Star Tribune*, October 16, 2019, https://www.startribune.com/disability-rights-advocate-sues-rental-scooter-companies-alleging-blocked-sidewalks/563240172/.
83. See, e.g., *Social Bicycles LLC d/b/a Jump v. City of L.A.*, Case No. 2:20-CV-02746, Complaint for Injunctive and Declaratory Relief (C.D. Cal. 2020).
84. Dave Colon, "Today: City to Announce Cargo Bike Delivery Pilot with Amazon, UPS, Others," *Streetsblog NYC*, December 3, 2019, https://nyc.streetsblog.org/2019/12/03/breaking-city-to-announce-cargo-bike-delivery-pilot-with-amazon-others/.
85. Angie Schmitt, "UPS to Test E-Bike Deliveries in Seattle," *Streetsblog USA*, October 30, 2018, https://usa.streetsblog.org/2018/10/30/ups-to-test-e-bike-deliveries-in-seattle/.
86. "LACI Launches First-in-Nation Zero Emissions Delivery Zone with City of Santa Monica and Partners Including Nissan, IKEA," City of Santa Monica, February 25, 2021, https://www.santamonica.gov/press/2021/02/25/laci-launches-first-in-nation-zero-emissions-delivery-zone-with-city-of-santa-monica-and-partners-including-nissan-ikea.

5. SCALING UP RENEWABLE ENERGY

1. "Ready for 100," Sierra Club, accessed March 16, 2022, https://www.sierraclub.org/ready-for-100.
2. U.S. Energy Information Administration, *International Energy Outlook 2021* (Washington, DC: U.S. Department of Energy, October 6, 2021), 1, https://www.eia.gov/outlooks/ieo/pdf/IEO2021_Narrative.pdf.
3. James H. Williams, Benjamin Haley, and Ryan Jones, *Policy Implications of Deep Decarbonization in the United States*, a report of the Deep Decarbonization Pathways Project of the Sustainable Development Solutions Network and the Institute for Sustainable Development and International Relations (San Francisco: Energy and Environmental

Economics, Inc., November 17, 2015), https://irp.cdn-website.com/be6d1d56/files/uploaded/2015-report-on-policy-implications.pdf.

4. Heather House and Lacey Shaver, "Beyond Buying Renewables: How Cities Can Influence the Energy System," Rocky Mountain Institute, July 27, 2020, https://rmi.org/beyond-buying-renewables-how-cities-can-influence-the-energy-system/.
5. Ariel Drehobl and Lauren Ross, *Lifting the High Energy Burden in America's Largest Cities: How Energy Efficiency Can Improve Low Income and Underserved Communities* (Washington, DC: American Council for an Energy-Efficient Economy, April 2016), 5, https://www.aceee.org/sites/default/files/publications/researchreports/u1602.pdf.
6. See, e.g., Adrian Wilson, Jacqui Patterson, Kimberly Wasserman, Amanda Starbuck, Annie Sartor, Judy Hatcher, John Fleming, and Katie Fink, *Coal Blooded: Putting Profits Before People* (Baltimore, MD; Bemidji, MN; Chicago: National Association for the Advancement of Colored People, Indigenous Environmental Network, Little Village Environmental Justice Organization, 2014), https://naacp.org/resources/coal-blooded-putting-profits-people.
7. "Local Government Climate and Energy Goals," American Council for an Energy-Efficient Economy, accessed July 29, 2020, https://database.aceee.org/city/local-government-energy-efficiency-goals.
8. City of Ann Arbor, Michigan, *A2Zero: Ann Arbor's Living Carbon Neutrality Plan* (Ann Arbor, MI: April 2020), https://www.a2gov.org/departments/sustainability/documents/a2zero%20climate%20action%20plan%20_3.0.pdf.
9. "Ann Arbor City Council Adopts A2Zero Carbon Neutrality Plan," City of Ann Arbor, Michigan, June 2, 2020, https://www.a2gov.org/news/pages/article.aspx?i=694.
10. "Utility Green Tariffs," World Resources Institute, accessed September 1, 2020, https://www.wri.org/our-work/project/clean-energy/utility-green-tariffs.
11. "Procurement Guidance: Green Tariff," American Cities Climate Challenge, accessed September 1, 2020, https://cityrenewables.org/green-tariffs/.
12. "Procurement Guidance: Green Tariff."
13. "Procurement Guidance: Off-Site Physical PPA," American Cities Climate Challenge, accessed August 7, 2020, https://cityrenewables.org/off-site-physical-ppa/.

14. 18 C.F.R. § 35, subpt. H.
15. "Procurement Guidance: Off-Site Physical PPA."
16. Eric O'Shaughnessy, Jenny Heeter, Julien Gattaciecca, Jenny Sauer, Kelly Trumbull, and Emily Chen, *Community Choice Aggregation: Challenges, Opportunities, and Impacts on Renewable Energy Markets*, NREL/TP-6A20-72195 (Golden, CO: National Renewable Energy Laboratory, February 2019), iv, https://www.nrel.gov/docs/fy19osti/72195.pdf.
17. Kaitlyn Bunker, Stephen Doig, Erik Fowler, James Mandel, and Christa Owens Michelet, *Community Energy Resource Guide* (Boulder, CO: Rocky Mountain Institute, December 2015), https://rmi.org/wp-content/uploads/2017/04/Community_Energy_Resource_Guide_Report_2015.pdf.
18. O'Shaughnessy et al., *Community Choice Aggregation*, iv.
19. "CCA by State," Lean Energy US, accessed August 10, 2022, https://leanenergyus.org/cca-by-state/.
20. O'Shaughnessy et al., *Community Choice Aggregation*, iv.
21. "Hudson Valley Community Power," Hudson Valley Community Power, accessed August 10, 2022, https://www.hudsonvalleycommunitypower.com.
22. "CCA by State," Lean Energy US, accessed March 16, 2022, https://www.leanenergyus.org/cca-by-state.
23. California Public Utilities Commission, *Power Charge Indifference Adjustment* (San Francisco: California Public Utilities Commission, January 2017), 1, https://www.cpuc.ca.gov/uploadedfiles/cpuc_public_website/content/news_room/fact_sheets/english/pciafactsheet010917.pdf.
24. Jeff St. John, "California to Hike Fees for Community Choice Aggregators, Direct Access Providers," *Greentech Media*, October 11, 2018, https://www.greentechmedia.com/articles/read/california-to-hike-fees-on-community-choice-aggregators-direct-access.
25. For more about terms used in a variety of franchise agreements, see "Municipal Franchise Agreements and Energy Objectives," National Renewable Energy Laboratory Data Catalog, December 4, 2019, https://data.nrel.gov/submissions/124.
26. "Municipal Franchise Agreements."
27. See, e.g., City of Arvada, Colo. Code Div. 2 § 7.2 (2015). Franchise agreements are often codified in the municipal code.

28. See, e.g., Franchise Agreement between the Town of Eaton, Colo. and Public Service Corp. of Colo. § 14.1 (2018).
29. *Memorandum of Understanding: Clean Energy Partnership*, City of Minneapolis and Northern States Power Company d/b/a Xcel Energy (October 2014), https://cleanenergypartnership.files.wordpress.com/2014/12/xcel-mou.pdf; *Memorandum of Understanding: Clean Energy Partnership*, City of Minneapolis and CenterPoint Energy (October 2014), https://mplscleanenergypartnership.org/wp-content/uploads/2014/12/centerpoint-mou.pdf.
30. City of Minneapolis, Xcel Energy, and CenterPoint Energy, *Minneapolis Clean Energy Partnership: 2019–2021 Work Plan* (Minneapolis, MN: City of Minneapolis, Xcel Energy, CenterPoint Energy, November 8, 2018), 6, https://mplscleanenergypartnership.org/wp-content/uploads/2018/11/CEP-2019-2021-Work-Plan_FINAL-APPROVED.pdf.
31. *Salt Lake City Corporation and Rocky Mountain Power Joint Clean Energy Cooperation Statement* (August 2016), http://www.slcdocs.com/slcgreen/Climate%20&%20Energy/CooperationStatement.pdf.
32. *Memorandum of Understanding Between the City of Charlotte and Duke Energy Carolinas to Establish a Low Carbon, Smart City Collaboration* (January 16, 2019), https://charlottenc.gov/sustainability/seap/SEAP/Duke%20MOU.PDF.
33. *Memorandum of Understanding Between the City of Charlotte and Duke Energy Carolinas*, 3–4.
34. *Memorandum of Understanding Between the City of Charlotte and Duke Energy Carolinas*, 4.
35. *Renewable Energy, Energy Efficiency, and Sustainability Agreement Between the City of Sarasota, Florida, and Florida Power & Light Company* (November 2010), https://www.sarasotafl.gov/home/showdocument?id=1008; *Memorandum of Understanding Between the City of Madison and MGE Regarding a Framework for Collaboration on Shared Energy Goals* (September 2017, https://madison.legistar.com/View.ashx?M=F&ID=5368330&GUID=A2A66EE0-94AA-4387-A24F-D8A0F9D1494D); *Energy Future Collaboration: Memorandum of Understanding Between the City and County of Denver, Colorado and Xcel Energy* (February 2018), https://www.denvergov.org/content/dam/denvergov/Portals/771/documents/EQ/DenverXcelMOU.pdf; *Memorandum of Understanding Between the City of Charlotte and Duke Energy Carolinas.*

36. Leah Cardamore Stokes, *Short Circuiting Policy: Interest Groups and the Battle Over Clean Energy and Climate Policy in the American States* (New York: Oxford University Press, 2020), 106.
37. "Green Power," Austin Energy, accessed September 1, 2020, https://austinenergy.com/ae/green-power.
38. "Renewable Energy," Austin Energy, accessed September 1, 2020, https://austinenergy.com/ae/green-power/renewable-energy.
39. "Where Does Your Electricity Come From?," Jefferson County Public Utility District, accessed September 1, 2020, https://www.jeffpud.org/fuel-mix/.
40. Bonneville Project Act, 16 U.S.C. § 832c (1937).
41. See generally Abby Briggerman, Radu Costinescu, and Ashley Bond on behalf of the American Public Power Association, *Survey of State Municipalization Laws* (Washington, DC: Duncan & Allen, May 2012), https://www.publicpower.org/system/files/documents/muncipalization-survey_of_state_laws.pdf.
42. Briggerman et al., *Survey of State Municipalization Laws.*
43. Paul Hughes, *Renegotiating a Municipal Franchise During Electricity Restructuring and Deregulation*, prepared for the American Public Power Association (Fairfax, VA: Environmental Services, Inc., July 2002), 10, http://www.informedcynic.com/SEC/buyout-docs/Renegotiating%20a%20Franchise.pdf.
44. Briggerman et al., *Survey of State Municipalization Laws.*
45. FERC Order No. 888 (1996).
46. Briggerman et al., *Survey of State Municipalization Laws.*
47. See, e.g., Winter Park, Florida, where the local IOU reportedly spent more than $500,000 on a campaign opposing municipalization. Kevin Ridder, "Forming a Municipal Utility," *Appalachian Voice*, December 19, 2018, https://appvoices.org/2018/12/19/forming-a-municipal-utility/.
48. Briggerman et al., *Survey of State Municipalization Laws.*
49. Suedeen G. Kelly, "Municipalization of Electricity: The Allure of Lower Rates for Bright Lights in Big Cities," *Natural Resources Journal* 37, no. 1 (1997): 43, 54.
50. Pieter Gagnon, Robert Margolis, Jennifer Melius, Caleb Phillips, and Ryan Elmore, *Rooftop Solar Photovoltaic Technical Potential in the United States: A Detailed Assessment*, NREL/TP-6A20-65298 (Golden, CO: National Renewable Energy Laboratory, January 2016), 34.

51. Brad Plumer, "Solar Power Is Contagious: Installing Panels Often Means Your Neighbors Will Too," *Vox*, April 1, 2015, https://www.vox.com/2014/10/24/7059995/solar-power-is-contagious-neighbor-effects-panels-installation.
52. Aurora Ord. No. 2011–05, codified as City Code of Aurora, Colo. § 146–1280 et seq. (2011).
53. Austin, Minn. Code of Ordinances § 11.84 (2013).
54. Town of Watertown Zoning Ord. art. VIII § 8.05 (2018).
55. N.Y. City Local L. 94 (2019).
56. Santa Monica Muni. Code. §§ 8.36, 8.106 (2017).
57. Cal. Bldg. Energy Efficiency Stds. for Residential and Nonresidential Buildings § 150.1(c)(14) (2019).
58. Seattle Energy Code §§ C411, C412 (2015).
59. Tucson Solar Ready Ordinance, Ord. No. 10549 (2008).
60. 780 C.M.R. ch. 115AA (2016).
61. City of Boulder Residential Energy Code § R406 (2020).
62. City of Boulder Residential Energy Code §§ R403.9, R403.10 (2020).
63. K. K. DuVivier, "Distributed Renewable Energy," in *Legal Pathways to Deep Decarbonization in the United States*, ed. Michael B. Gerrard and John C. Dernbach (Washington, DC: Environmental Law Institute, 2019), 517.
64. DuVivier, "Distributed Renewable Energy," 497.
65. DuVivier, "Distributed Renewable Energy," 498.
66. DuVivier, "Distributed Renewable Energy," 498.
67. DuVivier, "Distributed Renewable Energy," 498.
68. *City of Boise SolSmart Solar Statement* (Boise, ID: Office of the Mayor, Lauren McLean, July 13, 2020), https://www.cityofboise.org/media/3682/solsmart-solar-statement-signed-2-5-18.pdf.
69. DuVivier, "Distributed Renewable Energy," 498.
70. See, e.g., "Solar Installations," City of Aurora, Colorado, accessed July 30, 2020, https://www.auroragov.org/cms/one.aspx?pageId=16529398. For a range of local laws aimed at facilitating the development of distributed solar energy, see "Model Laws for Deep Decarbonization in the United States," available at https://lpdd.org.
71. "Community Solar for Home," NY-Sun (Solar Program), New York State Energy and Research Development Authority, accessed July 30, 2020, https://www.nyserda.ny.gov/All-Programs/Programs/NY-Sun/Solar-for-Your-Home/Community-Solar.

72. "Community Solar for Home."
73. Timothy DenHerder-Thomas, Jonathan Welle, John Farrell, and Maria McCoy, *Equitable Community Solar: Policy and Program Guidance for Community Solar Programs That Promote Racial and Economic Equity* (Washington, DC: Institute for Local Self-Reliance, February 2020), https://ilsr.org/wp-content/uploads/2020/02/Equitable-Community-Solar-Report.pdf.
74. "Community Solar," Solar Energy Industries Association, accessed August 10, 2022, https://www.seia.org/initiatives/community-solar.
75. Sara E. Bergan, Andrew P. Moratzka, Sarah Johnson Phillips, and David T. Quinby, "Community Solar," in *The Law of Solar: A Guide to Business and Legal Issues* (Portland, OR: Stoel Rives LLP,6th ed., 2017), chapter 9, 1–3, http://files.stoel.com/files/books/LawofSolar.PDF.
76. "How to Install Solar Panels on City-Owned Property and Lead by Example," C40 Knowledge Hub, March 2019, https://www.c40knowledgehub.org/s/article/How-to-install-solar-panels-on-city-owned-property-and-lead-by-example?language=en_US.
77. DuVivier, "Distributed Renewable Energy," 494.
78. See generally DuVivier, "Distributed Renewable Energy," 493.
79. Adele Peters, "Electric School Buses Are an Ingenious Solution to Help Utilities Build More Battery Storage," *Fast Company*, December 2, 2019, https://www.fastcompany.com/90436347/electric-school-buses-are-an-ingenious-solution-to-help-utilities-build-more-battery-storage.
80. Joann Muller, "Electric School Buses Are Batteries for the Grid," *Axios*, January 10, 2020, https://www.axios.com/electric-school-buses-vehicle-to-grid-power-19f7b6b1-662b-4501-a96e-dcf3fd57a886.html.
81. Dan Welch, *Electrified Transportation for All: How Electrification Can Benefit Low-Income Communities* (Arlington, VA: Center for Climate and Energy Solutions, November 2017), https://www.c2es.org/site/assets/uploads/2017/11/electrified-transportation-for-all-11-17-1.pdf.
82. Robert Lasseter, "CERTS Microgrid Concept," Consortium for Electric Reliability Technology Solutions, accessed September 3, 2020, https://certs.lbl.gov/initiatives/certs-microgrid-concept.
83. Erica Gies, "Microgrids Keep These Cities Running When the Power Goes Out," *Inside Climate News*, December 4, 2017, https://insideclimatenews.org/news/04122017/microgrid-emergency-power-backup-renewable-energy-cities-electric-grid.

84. Doug Vine and Amy Morsch, *Microgrids: What Every City Should Know* (Arlington, VA: Center for Climate and Energy Solutions, June 2017), 3, https://www.c2es.org/site/assets/uploads/2017/06/microgrids-what-every-city-should-know.pdf.
85. Vine and Morsch, *Microgrids*, 4.
86. Travis Sheehan, "Developing Smarter Cities: District Energy and Microgrids," *Smart Cities Dive*, accessed July 30, 2020, https://www.smartcitiesdive.com/ex/sustainablecitiescollective/how-cities-can-develop-smarter-district-energy-and-microgrids/153461/.
87. Hannah J. Wiseman, *Urban Energy*, 40 FORDHAM URB. L. J. 1809 (2013).
88. *Id.* at 1811.
89. *Id.* at 1811–12.
90. *Zimmerman v. Board of County Com'rs*, 289 Kan. 926 (Kan Supr. Ct. 2009).
91. Wiseman at 1816.
92. Wiseman at 1793, 1798.
93. See, e.g., Bethany Speer, Mackay Miller, Walter Schaffer, Leyla Gueran, Albrecht Reuter, Bonnie Jang, and Karin Widegren, *The Role of Smart Grids in Integrating Renewable Energy*, NREL/TP-6A20-63919 (Golden, CO: National Renewable Energy Laboratory, May 2015), https://www.nrel.gov/docs/fy15osti/63919.pdf.
94. *Naperville Smart Meter Awareness v. City of Naperville*, 900 F.3d 521, 528–29 (7th Cir. 2018).
95. Susan P. Altman, "Are You Ready for the Smart Grid?," K&L Gates, February 2, 2010, http://www.middletonslawyers.com/are-you-ready-for-the-smart-grid-02-01-2010/.

6. DECARBONIZING A CITY'S WASTE

1. Wee Kean Fong and Michael Doust, *Global Protocol for Community-Scale Greenhouse Gas Emission Inventories: An Accounting and Reporting Standard for Cities* (Washington, DC: World Resources Institute, 2014), 85–103.
2. *Philadelphia v. New Jersey*, 437 U.S. 617 (1978).
3. *C&A Carbone, Inc. v. Town of Clarkstown, N.Y.*, 511 U.S. 383, 392 (1994).
4. *United Haulers Ass'n v. Oneida-Herkimer Solid Waste Mgmt. Auth.*, 550 U.S. 330, 334 (2007).

5. *Id.* at 347.
6. Conn. G.S. §§ 22a-208(a), (b) (1986).
7. *Conn. Resources Recovery Authority v. Planning and Zoning Comm'n*, 225 Conn. 731, 754–55 (Conn. Supr. Ct. 1993).
8. 35 Pa. Stats. §§ 6018.101 et seq. (1980).
9. *Municipality of Monroeville v. Chambers Development Corp.*, 88 Pa. Commw. 603 (1985).
10. 53 Pa. Stats. §§ 4000.304(b)(2) et seq.
11. Ga. Code § 12-8-24 (2013).
12. Katie Sandson and Emily Broad Leib, *Bans and Beyond: Designing and Implementing Organic Waste Bans and Mandatory Organics Recycling Laws* (Cambridge, MA: Harvard Law School Food Law and Policy Clinic, July 2019), 54, citing the EPA Office of Solid Waste, Flow Controls and Municipal Solid Waste ES-1 (1995).
13. Sandson and Leib, *Bans and Beyond*, 54, citing Nicola Costantino and Roberta Pellegrino, "Evaluating Risk in Put-or-Pay Contracts: An Application in Risk Management Using Fuzzy Delphi Metric," *Journal of Applied Operational Research* 2 (2010): 62, 63.
14. Sandson and Leib, *Bans and Beyond*, 54.
15. See, e.g., Washington, D.C. Sustainable Solid Waste Mgmt. Amendment Act of 2014, D.C. Code ch. 10A (80 percent waste diversion requirement); San Francisco, Cal. Res. No. 679–02 (200) (zero-waste goal).
16. The climate impact of plastic bag bans has been questioned, particularly when the bags are replaced by single-use paper bags, which decompose more easily but are more carbon-intensive to produce and ship. Environment Agency, *Life Cycle Assessment of Supermarket Carrier Bags: A Review of the Bags Available in 2006*, report SC030148 (Bristol, UK: Environment Agency, 2011), https://assets.publishing.service.gov.uk/government/uploads/system/uploads/attachment_data/file/291023/scho0711buan-e-e.pdf.
17. Ariz. Rev. Stat. § 9–500.38 (2016); Colo. Rev. Stat. § 25-17-104 (1993); Idaho Code § 67–2340 (2016); Iowa Code § 331.301 (2020); Mich. S.B. No. 853 (2016); Minn. Stat. § 471.9998 (2019); Miss. S.B. 2570 (2018); Mo. Rev. Stat. § 260.283 (2015); N.D. Century Code § 23.1-08-06.1 (2019); Okl. Stat. § 2-11-504 (2019); S.D. S.B. 54 (2020); Tenn. Code § 7-51-2002 (2019); Tex. Health and Safety Code § 361.0961 (1993); Wis. Stat. § 66.0419 (2016). State statutes aggregated at "Defend Your Local Right to Reduce Plastic Pollution," Surfrider Foundation,

accessed August 10, 2022, https://www.surfrider.org/pages/defend-your-local-right-to-reduce-plastic-pollution.

18. N.Y. Env. Cons. L. §§ 27–2701 et seq. (2022).
19. *State Plastic and Paper Bag Legislation*, Nat'l Conf. of State Legislatures (Sept. 29, 2020).
20. Washington, DC's paper bag fee aims to address environmental injustice by directing proceeds to the cleanup of the Anacostia River. Revenue from New York State's paper bag fee, for the counties that impose it, is to be split between purchasing and distributing reusable bags to members of low- and fixed-income communities and the state's environmental protection fund. N.Y. State. Fin. L. § 92-S.
21. Jennie R. Romer and Leslie Mintz Tamminen, *Plastic Bag Reduction Ordinances: New York City's Proposed Charge on All Carryout Bags as a Model for U.S. Cities*, 27 TULANE ENVT'L L. REV. 237 (2014).
22. *Save the Plastic Bag Coal. v. City of Manhattan Beach*, 254 P.3d 1005 (Cal. 2011).
23. *Id.* at 246; e.g., *Schmeer v. Cnty of Los Angeles*, 153 Cal. Rptr. 3d 352, 354–55 (Cal. Ct. App. 2013). The charge was not held to be an unauthorized tax, in part because the money collected stayed with the retailer.
24. N.Y. Envt'l Conservation L. §§ 27–2801 et seq. (2022).
25. N.Y. Envt'l Conservation L. §§ 27–2701 et seq. (2022).
26. *Poly-Pak Industries, Inc. v. State of N.Y.*, Index No. 02673–20 (N.Y. Supr. Ct., Albany Cnty. 2020), Verified Article 28 and Declaratory Judgment Petition, Feb. 28, 2020.
27. *Poly-Pak Industries, Inc. v. State of N.Y.*, Index No. 02673–20 (N.Y. Supr. Ct., Albany Cnty. 2020), Decision/Order/Judgment at 39.
28. Ariz. Rev. Stat. § 9–500.38 (2016).
29. Colo. Rev. Stat. § 25-17-104 (1993).
30. *Gameros v. City of San Diego*, Verified Petition for Writ of Mandate and Complaint for Injunctive and Declaratory Relief, Docket No. 37-2019-00013383CU-TT-CTL (Cal. Super. Ct. 2019).
31. San Diego Muni. Code §66.0901 (2023).
32. *Restaurant Action Alliance v. Garcia*, Verified Petition (N.Y. Supr. Ct., N.Y. Cty. 2015).
33. *Restaurant Action Alliance v. Garcia*, Decision and Order, Index No. 100734/2015 (N.Y. Supr. Ct., N.Y. Cty. 2017), aff'd *Matter of Restaurant Action Alliance NYC v. City of New York*, 2018 NY Slip Op. 06975, 165 A.D.3d 515 (N.Y.A.D. 1st Dep't 2018).

34. E.g., Seattle, Wash. Muni. Code § 21.36.086 (2010) and Miami Beach, Fla. Code of Ordinances. § 46–213 (2019).
35. 42 U.S.C. §§ 12101–213 (2008).
36. See, e.g., Miami Beach Code of Ordinances. §46-213(b)(1) (2019).
37. Sandson and Leib, *Bans and Beyond*.
38. Sandson and Leib, *Bans and Beyond*.
39. See Sandson and Leib, *Bans and Beyond*, for a discussion of each of the following: Cal. Pub. Res. Code § 42649.81 (2018); Conn. Gen. Stat. Ann. § 22a-226e; 310 Mass. Code Regs. 19.017 (2022); N.Y. Envtl. Conservation Law (2019); R.I. Gen. Laws Ann. § 23–18.9-17 (2014); Vt. Stat. Ann. tit. 10, § 6605k (2011); Austin Code of Ordinances § 15-6-91 (2016); Boulder Mun. Code 6-3-13–18 (2019); Hennepin County Ordinance 13; Metro Code Ch. 5.10.410–470 (2018); N.Y.C. Admin. Code § 16–306.1 (2020); S.F. Env't Code §§ 1901–1912 (2009); Seattle Mun. Code 21.36.082–21.36.083 (2017).
40. *Bonesteel v. City of Seattle*, No. 15-2-17107–1 SEA, Order on Cross Motions for Summary Judgment at 11 (Wash. Superior Ct., King County 2016).
41. *Bonesteel v. City of Seattle* at 11.
42. San Francisco Ord. No. 100–09 (2009); San Francisco Env't Code §§ 1901 et seq. (2009); Peter Plastrik and John Cleveland, *Game Changers: Bold Actions by Cities to Accelerate Progress Toward Carbon Neutrality*, ed. Michael Shank and Johanna Partin (Copenhagen: Carbon Neutral Cities Alliance, 2018), 26.
43. Adam Redling, "Inside NYC's Organics Collection Program," *Waste Today*, November 26, 2019, https://www.wastetodaymagazine.com/news/new-york-city-organics-collection-waste-program/.
44. Hennepin County, Minn., Ordinance 13, § IV (Nov. 27, 2018).
45. Austin, Tex., ch. 15–16 Admin. Rules 8.5.1, 8.5.4 (2016).
46. Cal. Pub. Res. Code § 42652.5 (2018).
47. 30 Tex. Admin. Code ch. 332 (1995).
48. E.g., South Coast Air Qual. Mgmt. Dist. Rules 1133–1133.3 (2003). See also "Composting Emissions and Air Permits," California Department of Resources Recycling and Recovery (CalRecycle), accessed November 6, 2020, https://calrecycle.ca.gov/organics/air/.
49. E.g., 6 N.Y. Codes, Rules & Regs. § 361–2.1 et seq. (2021).
50. See, e.g., Zoning Ordinance of San Diego Cty. § 6912 (2022) and San Diego Municipal Code ch. 14, § 141.0820 (2022), which allows

community garden composting only if waste is limited to that of garden members or as long as amounts are less than five hundred cubic yards.

51. Sustainable Economies Law Center, *Growing Compost: A Policy Guide to Preserving Critical Community Composting in California* (Oakland, CA: Sustainable Economies Law Center, January 22, 2017), 16, https://d3n8a8pro7vhmx.cloudfront.net/theselc/pages/927/attachments/original/1485108714/Growing_Compost_Report_smaller.pdf?1485108714.
52. Fresno, Cal. Code of Ordinances. §§ 15–2750 (2018).
53. Madison Zoning Code Tbl. 27F-1 (2017).
54. Benjamin Miller, *Waste: Managing New York's Municipal Solid Waste to Support the City's Goal of Reducing Greenhouse Gases by 80 percent by 2050*, prepared for the New York League of Conservation Voters Education Fund (July 13, 2017), https://nylcvef.org/wp-content/uploads/2017/08/Waste-White-Paper.pdf, 20.
55. Los Angeles Ord. No. 182986 (2014); Los Angeles Muni. Code §§ 66.03, 66.33 et seq (2014).
56. N.Y. City Local L. 199 (2019).
57. Justin Wood, *Fighting New York's Climate Emergency with Waste Zones* (New York: Transform Don't Trash NYC, 2019), https://nylpi.org/wp-content/uploads/2019/06/Climate-CWZ-Report-Final-2019.pdf.
58. *Betz v. City of Los Angeles*, Case. No. BC664070, Cal. Super. Ct., Los Angeles, Complaint for Injunction (2017), referencing CAL. CONST. arts. XII C, D or Cal. Prop. 218 (1996).
59. N.Y. City Local L. 152 (2018).
60. *Nat'l Waste & Recycling Ass'n v. City of New York*, Index No. 101686/2018, Decision & Order on Motion at 3–4 (Oct. 3, 2019).
61. *Green Solutions Recycling v. Reno Disposal Co.*, 359 F. Supp. 3d 960 (2019); *Green Solutions Recycling v. Reno Disposal Co.*, No. 19–15201 (2020); *Nev. Recycling & Salvage v. Reno Disposal*, 134 Nev. Adv. Op. 55 (2018).
62. *Zolly v. City of Oakland*, Cal. Ct. of App. A154986 at 18 (2020).
63. National Solid Wastes Management Association, *Just Compensation and Hauler Displacement* (Washington, DC: National Solid Wastes Management Association, November 2012).
64. National Solid Wastes Management Association, *Just Compensation.*
65. *Ass'n of Taxicab Operators USA v. City of Dallas*, 720 F.3d 534, 541 (5th Cir. 2013). See also *Metro. Taxicab Bd. of Trade v. City of New York*, 615 F.3d 152, 155 (2d Cir. 2010).

66. "What Is a Circular Economy?," Ellen MacArthur Foundation, accessed September 14, 2020, https://ellenmacarthurfoundation.org/topics/circular-economy-introduction/overview.
67. Michael Burger, "Materials Consumption and Solid Waste," in *Legal Pathways to Deep Decarbonization in the United States*, ed. Michael B. Gerrard and John C. Dernbach (Washington, DC: Environmental Law Institute, 2019), 193.
68. "What Is a Circular Economy?"
69. C40 Cities, *Consumption-Based GHG Emissions of C40 Cities* (London: C40 Cities, March 2018), 4, https://cdn.locomotive.works/sites/5ab410c8a2f42204838f797e/content_entry5ab410fb74c4833febe6c81a/5ad4c0c274c4837def5d3b91/files/C40_GHGE-Report_040518.pdf?1540555698.
70. C40 Cities, *Consumption-Based GHG Emissions*, 9.
71. See ICLEI USA, *GHG Emissions in King County: 2017 Inventory Update, Contribution Analysis, and Wedge Analysis* (Denver, CO: ICLEI USA, July 2019), https://your.kingcounty.gov/dnrp/climate/documents/201907-KingCounty-GHG-Emissions-Analysis.pdf; Elizabeth A. Stanton, Ramon Bueno, Jeffrey Cegan, and Charles Munitz, *Consumption-Based Emissions Inventory for San Francisco* (Somerville, MA: Stockholm Environment Institute–U.S. Center, May 2011), https://sfenvironment.org/sites/default/files/fliers/files/sf_consumption_based_emissions_inventory.pdf.
72. See, e.g., Jonathan Rosenbloom, *Outsourced Emissions: Why Local Governments Should Track and Measure Consumption-Based Greenhouse Gases*, 92 U. COLO. L. REV. 451, 500 (2021).
73. Robert Sanders, "New Interactive Map Compares Carbon Footprints of Bay Area Neighborhoods," *Berkeley News*, January 6, 2016, https://news.berkeley.edu/2016/01/06/new-interactive-map-compares-carbon-footprints-of-bay-area-neighborhoods/.

INDEX

A2Zero Carbon Neutrality Plan, 98
accessory dwelling unit, viii, 38, 56
AFOLU. *See* agriculture, forestry and other land use
African Americans, 30
agriculture, forestry, and other land use (AFOLU), 7
air easements, 115
air pollution, 36, 63, 134
all-electric construction, viii, 50–52
ALPRs. *See* automated license plate readers
alternative pedestrian routes, 92
Americans with Disabilities Act (1990), 127
ANSI/ASHRAE/IES Standard 90.1, 43
anti-gift clause, 126
appliances: energy use standards for, 46–47; fossil fuel-powered, 46; regulation of, 42–43
area-wide pricing, 90
auditing, of energy, 59–60
Austin, Texas, 21–22, 67, 80, 104–5, 130–31
automated license plate readers (ALPRs), 89–90
autonomous vehicles (AVs), 85–86

Bean v. Southwestern Waste Management Corp, 24
benchmarking, viii, 59
benefit costs, 12
Berkeley, California, 47, 51, 54
bicycle infrastructure, 73, 91, 93
Biden, Joe, 27–28, 31
bikes, 19, 93–94
Bipartisan Infrastructure Law, 82
BIPOC, viii, 21, 36
Boulder, Colorado, 62
building code, viii; in compliance pathway, 49; decarbonization measures in, 44; energy efficiency in state, 43–44; EPCA exceptions to, 47–49; for EV charging, 80; local government's authority

building code (*continued*)
for, 44–45; local government without authority for, 49–50; low-carbon materials in, 69–70; for solar energy, 108–9; from state governments, 43–45, 62
building envelope, viii, 44
buildings: benchmarking laws for, 59; compliance pathway and, 62; decarbonization of existing, 66–67; decarbonization strategies for, 39–41, *40*; electrification of, x, 50–51; energy-efficient, 156n114; equitable policy for, 70–71; GHG emission policies for, 68–69; GHG emissions from existing, 58–59; GHG emissions of, 3–4, 39; GHG emission sources of, 42; land use and emissions from, 54–56; LEED standards for, 56–58; municipal, 112; noncompliance fees for, 63–64; PACE for, 67–68; performance standards for existing, 61–62; prescriptive requirements for, 65–66; regulation of, 41–42; rental, 62; retrofit's for existing, 41; retrofitting city, 35, 70–71; small, 64; state law policy toward, 35; state law preemptions for, 62–63; sustainability ratings of, 58; upgrading, 64–65; waste from operations of, 69–70; zoning codes for sizes of, 56; zoning energy codes for, 159n35; zoning requirements for, 55
CAA. *See* Clean Air Act
C&A Carbone, Inc. v. Town of Clarkstown, N.Y., 119
California Consumer Privacy Act, 89
California Environmental Quality Act, 127
California Restaurant Association (CRA), 54
carbon emissions, 103, 123
carbon-free energy, 140
carbon mitigation, 9, 97
carbon neutrality, 1
carbon pricing, 12
cargo e-bikes, 93–94
CCA. *See* community choice aggregation
charging networks: DC, 82–83; for EVs, 77–78; regulated as utilities, 79; in zoning codes, 84
circular economy, 137–38
cities. *See* local government
city buildings, 70
city government: GHG and energy consumption, 7–8; land use decisions of, 17–18; state law followed by, 14; taxes established by, 12
City of Richmond v. J. A. Croson, 23–24
civil rights, 25
Civil Rights Act (1964), 25, 28
clarify requirements, 81
Clean Air Act (CAA), 4, 15, 74–75, 136
clean energy, viii, 31; carbon mitigation for, 97; GHG

reductions from, 5; Inflation Reduction Act and, 29; PACE for, 67–68; renewable energy for, 103
climate change: energy-efficient buildings and, 156n114; laws not written for, 140–41; local government's action for, 141–42; plastic bags and, 170n16; policy-making for, 141–42; policy options on, 2
climate city, 20
climate justice, viii, 21–23
Climate Leadership and Community Protection Act (2019), 30–31
climate policy: decision-making on, 32; equitable, 31–32, 35–38; legal authority for, 141–42; race in, 22
Clinton, Bill, 27
commercial waste collectors, 133–36
Commercial Waste Zone Plan, 135
communities: disadvantaged, 20, 27–31; frontline, xi, 32, 34; Green, 50; landfills in, 118–19; local air pollution in, 20; low-income, 113, 118; solar energy for, 110–11; waste in gardens of, 172n50
community choice aggregation (CCA), viii, 101–2
Community Energy Partnership, 103
community gardens, 172n50
Complete Streets, 92
compliance pathway, ix, 44; building codes in, 49; building standards in, 62; prescriptive building requirements in, 65; solar energy in, 108
composting, 131–32, 172n50
compressed natural gas, 78
Computer Fraud and Abuse Act (1986), 60
conflict preemption, 11
congestion pricing, ix, 86; cities implementing, 13–14; privacy concerns in, 16; programs, 87–90; in transportation, 4, 72–73
Congress, on environmental justice, 28–29
Constitution, U.S.: Equal Protection Clause of, 23–24; federal government authority from, 75–76; Fifth Amendment of, 17, 41–42; Fourteenth Amendment of, 16, 24; Fourth Amendment of, 15–16; LTZs concerns of, 87–88; restrictions in, 15–16; Supremacy Clause of, 15
construction: all-electric, viii, 50–52; legal issues and requirements for, 3–4; waste from, 70
consumers, 83–84
consumption-based emissions, 7, 138–39
contracting, for green jobs, 32–33
cool roof, 66
cordon pricing, ix, 90
Court of Appeals for the Fourth Circuit, U.S., 30
Court of Appeals for the Ninth Circuit, U.S., 51

COVID-19 pandemic, 89, 130, 140
CRA. *See* California Restaurant Association
cultural equity, 21
cybersecurity, 60, 116

data: collection, 60–61; security, 60–61, 89, 116
DC. *See* direct-current rapid charger
DCC. *See* Dormant Commerce Clause
decarbonization, 35; building code measures for, 44; building strategies for, 39–41, *40*; of existing buildings, 66–67; financial burden of, 118; GHG emissions and, 69–70; municipal utilities electricity, 104–5; strategies, 3–4
decision-making, 17–18, 32
DEEP. *See* Department of Energy and Environmental Protection
de facto mandate, ix, 75, 86, 136
delivery trucks, 93–94
demand-response, ix, 40, 114
density bonus, ix, 81
Department of Energy and Environmental Protection (DEEP), 121
Department of Transportation, U.S., 28
deregulated (electricity), ix, 96
Dillon's Rule, ix, 10
direct-current (DC) rapid charger, ix, 82–83
direct requirements, 81
disadvantaged communities, 20, 27–31
disclosure policies, 59–60
discrimination, 25
disposable bags, 127
disposal and treatment of waste, 6
distributed energy generation, x, 96
distributed renewable energy, 106
distributed wind energy, 114–15
distributional equity, 21
Dormant Commerce Clause (DCC), x, 87, 119–20; out-of-state economic actors in, 15; transportation policies in, 76; on waste processing, 6
Drive Clean Seattle program, 79

electric readiness, x, 44
electric vehicles (EVs), x, 36, 72, 75; alternate modes of travel with, 91; building codes for charging of, 80; charging networks for, 77–78; local government's charging of, 82–83; municipal fleets of, 84; private property charging of, 80–81; public utility law on, 78–80; state governments and offenses involving, 83
electrification: energy and, 97; heat pumps and, 70; of heavy-duty vehicles, 85; utilities and, 78–79
electrification (building electrification), x, 50–51
Electronic Communications Privacy Act, 60
embodied carbon, x, 39–40, 69–70

eminent domain, x, 17
energy: audits of, 59–60; carbon-free, 140; city government and consumption of, 7–8; clean, viii, 5, 29, 31, 67–68, 97, 103; disclosure policies for, 59–60; distributed generation of, x, 96; distributed renewable, 106; distributed wind, 114–15; electricity and, 97; EPCA setting standards for, 46–47; equitable transition in, 36–37; equity in policies on, 97; financing for conservation of, 67–68; GHG emission reduction from, 58–59; GHG emissions from, 5, 57; government regulations of, 46–47; justice, x–xi; Justice40 for spending on, 27; local government authorities over, 95; local government costs of, 37; PPA for, 100; smart grid technologies for, 115–16; solar, 106–15; solar panels for, 106–7; storage of solar, 112–13; V2G, 113; waste-to-, xviii. *See also* green energy; renewable energy
energy-aligned leases, x, 68
energy efficiency, x, 103, 140; of buildings, 156n114; rental buildings with standards for, 62; in state building codes, 43–44
Energy Policy Act (1992), 43
Energy Policy and Conservation Act (EPCA, 1975), 4, 15, 43, 136; appliance energy use standards from, 46–47; building code exceptions to, 47–49; exceptions to, 47; local law preemption and, 54; ordinances preempted by, 51; transportation and, 74–75
Energy Rating Index, 109
enforcement mechanisms, 129
environmental justice, xi, 133, 171n20; climate justice and, viii, 21–23; Congress on, 28–29; federal law for, 27–28; laws for, 22–23; NEPA in, 26–27; state and local law on, 29–30; waste facilities in, 37
Environmental Protection Agency, U.S., 28
environmental review, xi, 26, 53, 77
EPCA. *See* Energy Policy and Conservation Act
EPS. *See* expanded polystyrene
equal protection, xi, 16, 23–24, 41, 141
Equal Protection Clause (1868), 16, 23–24
equitable policy: for building's GHG emissions, 70–71; on energy, 97; local laws with, 38; for lower-income communities, 118; state law barriers to, 33–34; in transportation, 73
equity: in climate policy, 31–32, 35–38; in energy policy, 97; energy transition with, 36–37; law on waste, 135; local air pollution and, 118; local law policies for, 38; race-neutral requirements for, 33; in transportation policy, 73; in transportation systems, 34; types of, 21; in waste management, 37
EVs. *See* electric vehicles

exclusive waste-carting franchise agreement, 131–32
Executive Order 14008 (2021), from Biden, 31
expanded polystyrene (EPS), 124, 126–27
extruded polystyrene, 127

federal-aid highway, xi
Federal Aviation Administration Authorization Act (1994, U.S.), 87
Federal Energy Regulatory Commission, U.S., 26, 96
federal government, U.S.: Bipartisan Infrastructure Law from, 82; highway eligibility funding from, 88; LTZs preemption from, 86–87; regulatory authority of, 42–43; U.S. Constitution authority to, 75–76
federal laws, U.S., 4; on civil rights, 25; on environmental justice, 27–28; municipalities limited by, 14; state and local law preempted by, 15, 74–75; on transportation, 74–75
fees, xi; as benefit costs, 12; on noncompliant buildings, 63–64; on paper bags, 171n20; parking in-lieu, 92; regulatory, 114n3; on waste, 122
field preemption, 11
Fifth Amendment, of U.S. Constitution, 17, 41–42
finance, 67–68, 118
fines, 12–13
fleet pricing, 90
flow control law, 119, 122
food donations, 130–31
for-hire vehicle, 90
fossil fuel–powered appliances, 46
fossil fuel–powered trucks, 133
Fourteenth Amendment, of U.S. Constitution, 16, 24
Fourth Amendment, of U.S. Constitution, 15–16
franchise agreement, xi, 53, 102–3
franchising programs, 133–34
freight vehicles, 85
frontline communities, xi, 32, 34
fuel economy, xii, 74–75
funding, state law limitations on, 33–34

Gallaher v. City of Santa Rosa, 152n45
gasoline sales, 83–84
gas stations, 84
gentrification, xii, 35, 64–65
GHG. *See* greenhouse gas
Global Protocol for Community-Scale Greenhouse Gas Emission Inventories (GPC), 3, 7, 117, 138
Global Protocol methodology, 138–39
government. *See* city government; federal government; local government; state government
GPC. *See Global Protocol for Community-Scale Greenhouse Gas Emission Inventories*
Green Communities, 50
green energy: jobs in, 32–33; leases for, xii, 68; policies on, 5; roofs

for, 66; tariffs on, xii, 99–100; zoning for, 107–8
greenhouse gas (GHG): buildings emission sources of, 42; buildings emissions policies for, 68–69; buildings emission standards on, 61–62; buildings with emissions of, 3–4, 39; city's energy consumption and, 7–8; clean energy reducing, 5; in climate goals, xii; consumer measures addressing, 83–84; consumption-based emissions and, 138–39; decarbonization and emissions of, 69–70; energy emissions of, 5; energy use and, 57; energy with reduction of, 58–59; equitable policy for emissions of, 70–71; from existing buildings, 58–59; local laws on, 18; reduction of, 123–24; renewable energy reducing, 95; state law authority on, 11; from transportation, 4, 72–74; from waste, 117–19; zoning requirements and, 38
green tariff, xii, 99–100

health, 30, 36
heating, ventilation, and air conditioning. *See* HVAC
heat pump, xii, 46, 55, 70, 97
heavy-duty vehicles, xii, 85
highways, funding for, 88
home rule petition, 50
Honolulu, equity types in, 21
hot-water heaters, 109
HVAC, xii, 42, 46
hybrid vehicles, xii, 75

independent system operator, xii, 96
Inflation Reduction Act (2022), 29
Infrastructure Investment and Jobs Act (2021), 29
intergenerational equity, 21
internal combustion engines, 36, 78
International Energy Conservation Code, 43
investor-owned utilities (IOUs), xiii, 105, 166n47

just compensation, xiii, 17–18, 82, 105
Justice40, 27

landfills, 118–19, 122, 133
land use, xiii; building emissions and, 54–56; legal issues in, 17–18; municipalities authority on, 91–92; in U.S., 37–38
last-mile delivery, 93–94
laws: building benchmarking, 59; for climate change, 140–41; for environmental justice, 22–23; flow control, 119, 122; New York climate, 29–30; preemption, 11–12; private, 18; on public utilities, xv, 63, 78–80; recycling, 126; state public utility, 34–35, 52; urban climate, 18; on waste equity, 135. *See also* federal laws; local laws; state laws
Leadership in Energy and Environmental Design (LEED), 55–58

legal authority: in all-electric construction, 50–51; in climate justice, 22–23; in climate policy-making, 141–42; in local government, 50–51; in microgrid development, 114; in waste reduction, 119–24
legal issues: in carbon-free energy, 140; climate change and, 140–41; in commercial waste zones, 134–35; in community solar projects, 111; construction requirements with, 3–4; in land use, 17–18; in pay as you throw, 128; smart meters with, 116; in wind energy, 114–15
legislative authority, 14
LEZs. *See* low-emission zone
license plate–reading cameras, 16
local (or localized) air pollution, xiii, 30, 133; in communities, 20; equity and, 118; from transportation, 36–37; from truck traffic, 85; from waste, 37, 123
local businesses, 15–16
local government: building codes authority of, 44–45; building codes without authority of, 49–50; building sustainability rating from, 58; climate action from, 141–42; congestion pricing implemented by, 13–14; energy costs of, 37; energy regulations from, 46–47; energy use authorities of, 95; equity definitions of, 21; EV charging from, 82–83; gasoline sales curbed by, 83–84; green zoning requirements from, 107–8; legal authority of, 50–51; market participant exceptions for, 76–77; penalties imposed by, 63–64; regulatory scenarios of, 79; road closings by, 89; third-party standards and, 57–58; utilities and, 5; utility agreements with, 103–4; waste legal obligations for, 122–23; zoning authority of, 51, 81, 115; zoning codes from, 54–55. *See also* communities; municipalities
Local Law 97 (2019, NYC), 65, 154n85, 155n100
Local Law 152 (2016, NYC), 135
local laws: on environmental justice, 29–30; EPCA and preemption of, 54; on EPS, 126–27; equitable policies from, 38; federal law preempting, 15, 74–75; on GHG, 18; micromobility regulations in, 19; on organic waste, 129–31; on plastic bags, 125–26; state law preempting, 10–12
low-carbon materials, 69–70
low-emission vehicles, 78
low-emission zones (LEZs), xiii, 72–73, 86
low-income communities, 113, 118
low-traffic zones (LTZs): constitutional concerns of, 87–88; federal preemption of, 86–87; road closings in, 89

mandatory recycling, 128–30
market participant exception, xiii, 76–77
Massachusetts, 45–46, 49–50
Massachusetts Stretch Energy Code, 108
megawatt, xiii, 96, 103
megawatt-hour (MWh), xiii
Michigan, 45, 56, 63
microgrids, xiii, 113–14
micromobility, 19, 93–94
mode shift, xiii, 19, 73, 91–92
Municipal Home Rule, xiv, 10
municipalities, 9; buildings in, 112; Dillon's Rule in, 10; federal laws limiting, 14; fines and penalties imposed by, 12–13; fleets in, 84; franchise agreements and, 102–3; green tariff's legal barriers for, 99–100; impact fees by, 144n3; land use authority by, 91–92; legislative authority traded away by, 14; microgrid development in, 114; road toll permission for, 13–14; from state law, 105–6; stretch codes from, 49; toll collecting by, 88; utilities in, xiv, 104–5; waste processing sites in, 132–33
municipalization, xiv, 34–35, 104–5
MWh. *See* megawatt-hour

National Environmental Policy Act (NEPA, 1970), 26–27, 30, 88
natural gas: bans on, xiv, 11; Berkeley's ban on, 54; compressed, 78; connections, 52; restrictions, 53
neighborhoods: gentrification of, 35, 64–65; racial segregation of, 37–38
net metering, xiv, 100, 106
net-zero emissions, 1, 98
net-zero stretch code, 46, 49–50
New York, 45; benchmarking data from, 59; climate laws in, 29–30; Climate Leadership and Community Protection Act of, 31; Local Law 97 in, 65; net-zero stretch code in, 49
nondelegation doctrine, xiv, 56–57

obligation to serve, 52
onboard payment mechanisms, 90
ordinances, xiv, 67, 83; EPCA preempting, 51; local flow control, 119, 122; zoning, 17, 107–9
organics-processing facilities, 132
organic waste, xiv, 129–31
out-of-state economic actors, 15
out-of-state waste, 120

PACE. *See* property-assessed clean energy
paper bags, fees on, 171n20
parking in-lieu fees, 92
parking minimums, 91–92
"pay as you throw," xiv, 128
pedestrian infrastructure, 73, 91–93
penalties, xiv; local government imposing, 63–64; municipalities imposing, 12–13; states approving, 12

Pennsylvania, 45, 121
performance standards (or performance pathway), xiv; for existing buildings, 61–62; on small buildings, 64
Philadelphia v. New Jersey, 119–20
Pike balancing test, 87
plastic bags, 124–26, 170n16
police powers, xiv–xv, 41, 51
policy: barriers to equitable, 33–34; building emissions, 68–69; carbon mitigation, 9; climate, 22, 31–32, 35–38, 141–42; climate change, 2; disclosure, 59–60; equitable, 33–34, 38, 70–71, 73, 97, 118; green energy, 5; state law building, 35; waste reduction, 124
power purchase agreement (PPA), xv, 100
preemption, xv; on all-electric construction, 52; for buildings, 62–63; conflict and field, 11; by federal laws, 15, 74–75; laws, 10–12; local law, 54; LTZs, 86–87; of ordinances, 51; from public utility law, 63; state, 52
prescriptive pathway, xv, 48
prescriptive requirements, xv, 65–66
privacy protection: congestion pricing concerns for, 16; data collection and, 60; in Fourth Amendment, 16; license plate-reading cameras and, 16; smart meters and, 116, 145n19
private laws, 18
private property, 80–81
private right of action, xv, 25
procedural equity, 21
procurement, xv, 32–33
property-assessed clean energy (PACE), xv, 67–68
public disclosure, 153n70
public roadways, 81–83
public transit, 34, 36, 73, 77
public trust, xv, 159n40
public utilities: charging networks regulated as, 79; data collection from, 61; EVs and laws of, 78–80; franchise agreement with, 53; laws and regulations on, xv, 63, 78–80; local government's agreements with, 103–4; renewable energy and, 98–99; state laws on, 52
public utility (or service) commission, xv; green tariffs and, 99–100; IOUs and, 106; retail sales regulated by, 96; TNCs and, 83
public utility (or service) law, xv, 4, 106; EVs and, 78–80; preemptive effect from, 63; state, 34–35, 52
"put or pay" contract, 122

race, 22, 33, 37–38
rate-making, xvi, 5, 99, 141
reach code, xvi
REC. *See* renewable energy certificate or credit

recycling: enforcement mechanisms on, 129; laws, 126; mandatory, 128–30; waste facilities for, 132
redlining, xvi, 71
regional transmission organization, xvi, 96
regulation, xvi; of appliances, 42–43; AVs requirements for, 85–86; of buildings, 41–42; of charging networks, 79; deregulated and, ix, 96; of energy, 46–47; federal government authority of, 42–43; fees, 114n3; micromobility, 19; of public utilities, xv, 63, 78–80; of retail sales, 96; of TNCs, 83
regulatory takings, xvi, 18
renewable energy, xvi; for clean energy, 103; distributed, 106; equity in energy policies for, 97; GHG emission reductions from, 95; goals set for, 98; green tariffs in, 99–100; microgrids for, 113–14; for municipal buildings, 112; public utilities and, 98–99; retrofits financing for, 67–68; utility-scale, 98–99
renewable energy certificate or credit (REC), xvi, 5
renewable portfolio standard, xvi, 105
Reno, Nevada, 62
rental buildings, 62
resilience, xvi, 96, 114
restrictions, in U.S. Constitution, 15–16
retail sales, 96
retrocommissioning requirement, 65–66
retrofit, xvi, 64, 102; for city buildings, 35, 70–71; for existing buildings, 41; PACE used for, 67; for renewable energy, 67–68; split-incentive problem for, 68
revenue, 12–13
right-of-way, of public roadways, 81–83
R.I.S.E. v. Kay, 24
road closings, 89
road tolls, 13–14
rooftop solar, 114–15

scooters, 19
setback, xvii, 55, 115
single-family zoning, 38
single-use paper bags, 170n16
single-use plastics, 127
small buildings, 64
smart grid technologies, 60, 115–16
smart meters, xvii, 16, 60–61, 116, 145n19
solar energy: building codes for, 108–9; for communities, 110–11; in compliance pathway, 108; distributed, 106–7; energy storage for, 112–13; panels for, 112; rooftop, 114–15; zoning codes for, 107–8
solar readiness, xvii, 44, 108
solid waste, 148n40
Solid Waste Management Act (1980, PA), 121

Solid Waste Management Plan, 148n40
source reduction, 123
split-incentive problem, xvii, 18, 68
state government: building codes from, 43–45, 62; climate action from, 141; energy regulations from, 46–47; environmental review by, 77; EVs charging offenses and, 83; penalties approved by, 12; preemption laws from, 11–12; public utility law from, 34–35, 52; road tolls permission required by, 13–14; toll collecting by, 88
state laws: on air pollution, 63; building policy in, 35; building's preemptions in, 62–63; CCA in, 101; cities following appropriate, 14; on environmental justice, 29–30; environmental review required by, 53; on EPS, 126–27; equitable policy barriers in, 33–34; federal law preempting, 15, 74–75; funding limitations in, 33–34; GHG and authority of, 11; home rule petition from, 50; local climate law preempted by, 10–12; municipalization from, 105–6; on plastic bags, 125; on public utilities, 52; on waste collection, 120–21
state preemption, 52
state public utility laws, 34–35, 52
statute, xvii, 15, 53, 82; environmental review, 76–77; heavy-duty vehicles and, 85; for road closures, 89
stranded costs, 105
stretch code, xvii, 49, 80
structural equity, 21
Supremacy Clause, of U.S. Constitution, 15
Supreme Court, U.S., 17, 76, 141; *Bean v. Southwestern Waste Management Corp* case from, 24; *City of Richmond v. J. A. Croson* case from, 23–24; *R.I.S.E. v. Kay* case from, 24; on Title VI, 25
sustainability ratings, 58

tailpipe pollutants, xvii, 72
Takings Clause, 18
tax, xvii, 12, 63–64, 122
taxi business, 83
third-party standards, 57–58
tiny houses, 56
Title VI, 25, 28
TNCs. *See* transportation network companies
toll collecting, 88–90
traditionally regulated (electricity), xvii, 96, 100
traffic, low-emission zones for, 72–73
transportation: congestion pricing in, 4, 72–73; DCC policies on, 76; environmental review of, 77; EPCA and, 74–75; equitable policy in, 73; equitable systems in, 34; federal laws concerning, 74–75; GHG emissions from, 4, 72–74; health influenced by, 36; local air pollution from, 36–37;

public transit network in, 36; of solid waste, 148n40; U.S. emissions from, 72
transportation network companies (TNCs), 83
trash hauling, 128, 136

United Haulers Association, Inc. v. Oneida-Herkimer Solid Waste Management Authority, 120
United States (U.S.): Court of Appeals for the Fourth Circuit in, 30; Court of Appeals for the Ninth Circuit in, 51; Department of Transportation in, 28; land use in, 37–38; mandatory recycling in, 128–29; transportation emissions in, 72
urban climate law, 18
U.S. *See* United States
utility, xvii; charging networks regulated as, 79; electrification and, 78–79; local government agreements with, 103–4; local government and, 5; municipalities and, xiv, 104–5. *See also* public utilities
utility-scale energy generation, xvii, 96

vehicle-to-grid (V2G), xviii, 113
virtual power purchase agreement (VPPA), xviii
Vision-Zero goal, 92
VPPA. *See* virtual power purchase agreement

waste: from building operations, 69–70; in circular economy, 137–38; commercial collectors of, 133–36; in community gardens, 172n50; from construction, 70; disposal and treatment of, 6; diversion, 123; equity in management of, 37; equity law, 135; exclusive waste-carting franchise agreement on, 131–32; fees on, 122; GHG emissions from, 117–19; legal frameworks for reducing, 119–24; legal issues in commercial, 134–35; local air pollution from, 123; local businesses refusing out-of-state, 15–16; local government's legal obligations on, 122–23; materials and product bans, 124; organic, xiv, 129–31; out-of-state, 120; plastic bags as, 124–26; policy reduction of, 124; recycling facilities for, 132; solid, 148n40; state laws on, 120–21; tax on, 122; trash hauling of, 128, 136; zero-waste goal for, xviii, 123, 132, 137–38; zoning requirements for, 121
waste-processing facilities, 37, 118, 132–33
waste-to-energy, xviii, 37, 121
wind energy, distributed, 114–15

zero-emission vehicles (ZEVs), xviii, 94, 133, 136
zero-waste, xviii, 123, 132, 137–38
ZEVs. *See* zero-emission vehicles

zoning, xviii; building energy codes in, 159n35; building requirements from, 55; GHG requirements for, 38; for green energy, 107–8; local government's authority in, 51, 81, 115; ordinances, 17, 107–9; requirements, 131; waste requirements of, 121

zoning codes: building sizes in, 56; charging networks in, 84; density bonuses in, 81; for gas stations, 84; from local government, 54–55; for microgrids, 114; for organics-processing facilities, 132; parking minimums in, 91–92; for solar energy, 107–8

Printed and bound by CPI Group (UK) Ltd, Croydon, CR0 4YY

12/11/2024

14591708-0001